ELECTRIC FIELD ENHANCED MEMBRANE SEPARATION SYSTEM: PRINCIPLES AND TYPICAL APPLICATIONS

ELECTRIC FIELD ENHANCED MEMBRANE SEPARATION SYSTEM: PRINCIPLES AND TYPICAL APPLICATIONS

SIRSHENDU DE,

BISWAJIT SARKAR

AND

SUNANDO DASGUPTA

Nova Science Publishers, Inc.

New York

For permission to use material from this book please contact us:
Telephone 631-231-7269; Fax 631-231-8175
Web Site: http://www.novapublishers.com

LIBRARY OF CONGRESS CATALOGING-IN-PUBLICATION DATA

De, Sirshendu.
Electric field enhanced membrane separation system : principles and typical applications / Sirshendu De, Biswajit Sarkar, Sunando DasGupta.
p. cm.
Includes bibliographical references and index.
ISBN 978-1-60741-592-3 (hardcover : alk. paper)
1. Membrane separation. 2. Electric fields. I. Sarkar, Biswajit. II. DasGupta, Sunando. III. Title.
TP248.25.M46D4 2009
660'.28424--dc22

2009015620

Published by Nova Science Publishers, Inc. ✝ New York

CONTENTS

PREFACE

Membrane separation process has emerged as a powerful separation technique and has become an integral part of modern process industries. This has found wide applications in the field of biotechnology, beverage, dairy industry, clarification of fruit juice, waste water treatment etc. The advantages of these processes are their unique separation capabilities, low energy consumption, high throughput, and ease of scaling etc., compared to conventional separation processes. However, the application of membrane separation process has been limited by the presence of concentration polarization and membrane fouling. These lead to decline in permeate flux and change in retention behaviour.

Therefore, the ways to reduce concentration polarization and thereby enhancing the filtration rate are currently active areas of research in view of the increasing demand of membranes application to various industrial separation processes. To minimize the membrane fouling and concentration polarization, several techniques have been investigated. In this regard, application of external d.c. (direct current) electric field across the membrane is a promising method to reduce concentration polarization and fouling. In presence of externally applied d.c. electric field with appropriate polarity, charged solutes migrate away from the membrane surface due to electrophoresis. Hence, the concentration polarization and deposition thickness are thereby reduced resulting in an enhancement of permeate flux.

This book deals with an extremely specialized topic on electric field enhanced membrane separation. The relevant principles, technology and the theoretical details are elaborated in the book. It is to be emphasized that no such book dealing with electric field assisted membrane filtration exists currently. Therefore, this book obviously has significant advancement compared to existing books on membrane technology. This book can be used as one of the Texts for the course like "Novel Separation Processes" taught in Postgraduate level. Of course, this book can be an extremely useful reference book for the students and professionals in Chemical and Environmental Engineering. We believe this book will have two fold impacts. Firstly, its academic value is quite high; Secondly, it will have remarkable impact of scaling up such system in actual industrial scale from pilot plant data in an emerging area. Moreover, a book on such a topic does not exist today; the importance of it from an academic point of view is undoubtedly remarkable.

This proposed book presents detailed description of the effect of electric field to a number of typical applications, e.g. fruit juice clarification, separation and fractionation of protein solution, micellar enhanced ultrafiltration for dye removal. The results are analyzed in full detail. We believe this book is a first kind of its own in this field. Since the topic is an emerging area and most of the work presented has potential of field application, the possibility of this book being out of date is rare in near future.

Chapter 1

INTRODUCTION

With the rapid growth in the field of chemical, biotechnological, pharmaceutical, food processing as well as waste treating industries, development of large scale separation and purification processes has become indispensable. Conventional separation processes such as distillation, centrifugation, adsorption, extraction etc., are limited by thermodynamic equilibrium. In recent years, much attention has been given to look for innovative and effective separation techniques to overcome the disadvantages encountered in the conventional separation processes. In this regard, the rate governed membrane based separation processes have gradually become an attractive alternative to many industrial applications. The advantages of these processes are low energy requirement, unique separation capabilities, easy to scale-up etc. Membrane is a barrier through which components are transported under the driving force of a gradient in their electrochemical potential. Gradients in the electrochemical potential of a component in the membrane interphase may be caused by differences in the hydrostatic pressure, the concentration, the temperature, or the electrical potential between the two phases separated by the membrane. The basic membrane systems commercially available are microfiltration (MF), ultrafiltration (UF), nanofiltration (NF) and reverse osmosis (RO) each of which is treated with a different range of particle size. The range of operating pressure, membrane pore size and molecular weight of the separating solutes are as follows:

Microfiltration (MF): This involves the use of membrane in the micron range. Operating pressure: 2 to 4 atmosphere; pore size: 0.1 to 5 micron. Ultrafiltration (UF): Ultrafiltration is designed to separate macromolecules such as proteins, starch, clays, paints, pigments etc. and performed at the pressure range of 3 to 8 atmosphere, a pore size of 10 to 200 Å and with a molecular weight of the separating solutes in the range of 1,000 to 5,00,000. Nanofiltration (NF): This is typically used to retain compounds such as sugars, divalent salts, dissociated acids etc. Operating pressure: 10 to 30 atmospheres; pore size: 10 to 20 Å; molecular weight of the retained solutes: 100 to 1000.

Reverse osmosis (RO): RO membrane is characterized by 95% retention of common salt, sodium chloride. Operating pressure: 35 to 100 atmospheres; pore size is 2 to 10 Å.

1.1. Application of Membrane Separation Processes

Pressure driven membrane separation processes are extensively used for a wide range of separation in diversified fields as chemical processing, water treatment, textile, dairy, food processing, biotechnology, pharmaceutical, pulp and paper, sugar, tannery. Some major applications of membrane processes are given below.

1.1.1. Electro-painting

One of the oldest applications of ultrafiltration in the chemical and mechanical industries is the recovery of paint in the electro-deposition painting process which is started about 30 years ago in automotive industry. During electro-deposition process, the paint which is washed off not only results an economic loss, it also causes a water pollution problem. The use of ultrafiltration largely avoids the loss of paints and minimizes the water pollution. Successful application of ultrafiltration has also been observed for the recovery of other coating materials such as paper coating, latex and flexo-graphic ink [1].

1.1.2. Water Purification

The effective removal of organic compounds has always been a major challenge for the production of potable water which must be essentially free from organics in order to be fit for human consumption. The major problems are due to excess of fluoride content, higher salinity level, total dissolved solid, hardness, heavy metals, color, organic compounds, e.g., total organic carbon, biological oxygen demand, chemical oxygen demand, total organic halides, trihalomethanes in underground water. Conventional water treatment technologies include coagulation, softening, activated carbon, ion exchange, oxidation (e.g., chlorination and ozonation). Membrane separation is a viable option for treatment of water and waste water by complying with the increasingly strict legislation concerning potable water quality and allowable wastewater discharge limit worldwide. Membrane separation processes have been progressively used for water and wastewater treatment in order to remove suspended solids and reduce the content of organic and inorganic matters [2,3]. Pervov et al. [4] reported that membrane separation is more advantageous than conventional treatment for the treatment of ground water containing excessive hardness, metal ions and total dissolved solids.

1.1.3. Textile Industry

The removal of dyes from industrial effluent is a major concern of the process industries like textile, paper, printing, cosmetics etc. These industries are major consumer of water and therefore cause water pollution. Treatment of waste water containing dyes is one of the most important ecological problems because the effluents containing dyes are not only rich in color but also toxic to aquatic life. Parameters of major interest in the textile wastewaters concern chemical oxygen demand (COD), biological oxygen demand (BOD), total organic carbon (TOC), dissolved solids, conductivity and color. Due to increasingly stringent environmental legislation, development of effective waste water treatment technologies is essential. Several techniques have been used to remove color from waste water. The most common are biological treatment [5], coagulation/ flocculation [6], oxidation [7], adsorption [8] etc. However, the efficacy of these treatments is not sufficiently high since most compounds cannot be easily degraded. In this regard, membrane filtrations are becoming attractive for the separation of colored effluent containing various types of dyes. Reverse osmosis and nanofiltration are widely used for the separation of several commercial dyes [9-12].

1.1.4. Dairy Industry

Membrane systems are used extensively throughout the dairy industry to control protein, fat and lactose contents of a variety of products [13]. Successful application of membrane separation technology in the processing of cheese whey has started in 1972. Whey, bi-product of the cheese industry, is the liquid fraction remaining after the recovery of cheese. Whey contains protein (20% of total protein), lactose, minerals and water-soluble vitamins. Since the cheese consumption is increasing around the world day by day, disposal of whey is a real problem. Hence use of membrane is appropriate to fractionate, purify and concentrate whey components, thereby increasing their utilization and reducing the pollution problem. Ultrafiltration is used in the cheese industry to fractionate the proteins from whey and to make cheese from ultrafiltered milk [14]. Microfiltration are generally used for the removal of lipid, microorganism prior to ultrafiltration, thus reducing the bio-burden without need for high temperature pasteurization [15]. Nanofiltration membranes are suitable in dairy application for partial demineralization of whey due to its high permeability of salt and very low permeability of lactose, protein, urea etc.[16].

1.1.5. Soy Protein Extraction

Soy proteins are being widely used as functional ingredients. In traditional methods, the soy protein is extracted from soy flour with alcohol or alkali, heat treatment, precipitation or centrifugation. As a result, traditional soy protein concentrates sometimes have poor functional properties and some of the proteins are lost with the whey-like waste stream. The main objective in producing a soy concentrate is to remove the oligosaccharides and minerals from the defatted soy flour, thereby increasing its protein content from about 50% in the soy flour to around 70% in the concentrate. Ultrafiltration can be a useful technique to separate smaller molecular weight oligosaccharides from the larger proteins and fiber fraction present in soy flour. [17, 18]. Moreover, in ultrafiltration, there is no need for any chemicals and the whey proteins are the part of the final product. This results in higher yield and superior functional properties of the ultrafiltered soy product.

1.1.6. Fruit Juice Clarification

Clarification of fruit and vegetable juice has been one of the most successful applications of membrane technology. Ultrfiltration is widely used for clarification of various juices and becomes an integral part of modern juice processing industry [19]. In fruit juice, apart from low molecular weight flavor, aroma components, sugars, acid and salt, higher molecular weight components like protein, pectin, cellulose, hemicellulose, starch, and spoilage microorganisms are also present. These higher molecular weight components are responsible for post bottling haze formation. Hence, it needs clarification. Traditional methods of fruit juice clarification involve several batch operations that are both time- and labor-consuming and involve the use of large amount of fining agent, enzymes, and diatomaceous earth, etc. as well as heat treatment to destroy spoilage microorganism. Ultrafiltration has been investigated for the clarification of apple, pear, orange, lemon, kiwifruit, etc. [20-23]. During ultrafiltration larger species such as pectin, protein, cellulose, hemicellulose, starch, etc. are retained while smaller species such as sugar, salt, vitamins, aroma compounds permeates through the membrane. Therefore, microbial contamination in the fruit juice can be minimized. Hence, shelf life of the fruit juice increases. Moreover, using ultrafiltration, the use of thermal treatment and loss of volatile aroma compounds are avoided.

1.1.7. Biotechnological Application

Downstream processing is one of the most cost intensive parts in bio-processing industry due to the high complexity of bio-suspensions itself, the low concentration of the product and the stress sensitivity of the valuable target molecules. In order to improve the performance, process selectivity must be increased. Conventional techniques such as chromatography, affinity separation, and electrophoresis can produce small quantities of very pure proteins but these processes are very difficult to scale-up and their throughput is very low. In this regard, membrane separation process has emerged as a powerful bio-separation technique for large scale purification, separation and concentration of several bio-products such as therapeutic drugs, enzymes, hormones, antibodies etc. Ultrafiltration processes give a very high throughput of product and can be fine tuned to give high selectivity in protein fractionation [24-27].

1.1.8. Pulp and Paper Industry

The pulp and paper industry is a large water consumer and discharges effluents containing high inorganic and organic pollutants. The common conventional treatment techniques are aerobic and anaerobic treatment, lime and alum coagulation and precipitation, oxidation, ion exchange. For the treatment of paper and pulp effluent, ultrafiltration becomes an attractive method as most of the higher molecular weight compounds are retained by this membrane [28].The kraft process is the dominating chemical pulping process. The recovery of valuables chemicals from black liquor is an integral part of kraft pulping process. In a conventional kraft recovery process, black liquor is concentrated, incinerated and causticized to recover sodium hydroxide, the required chemical for alkaline pulping. This process is very cost intensive as large amount of water is consumed during concentration. In this regard, microfiltration and ultrafiltration [29] seem to be a viable unit operation to recover valuable chemicals from black liquor and will be attractive with regard to pollution. Ultrafiltration removes all the suspended colloidal materials. UF and RO have been investigated for the treatment of white liquor from kraft black liquor [30].

1.1.9. Sugar Processing Industry

Sugar processing is one of the most energy-intensive processes in the food and chemical industries. Membrane separation processes have the potential for considerable saving in operating costs by eliminating the current energy intensive unit operations like evaporation and distillation [31]. The potential applicability of membrane operations in both the cane and beet sugar industry are reviewed and discussed by several researchers. Ultrafiltration has been successful for the clarification as well as decolorization of raw brown sugar obtained from Indian sugar industry [32]. Clarification by UF of sugar solution is proven to be technically superior to ordinary lime treatment because it yields a juice of higher purity and better color quality and is free from starch and acidified substances [33]. Moreover, ultrafiltration of sugar juice makes the crystallization process easier than in the juices conventionally purified due to better separation of non-sucrose compounds and lowering its viscosity. Madaeni et al. [34] have applied a two-stage reverse osmosis system for pre-concentration of thin sugar juice and showed a significant energy saving of about 33% compared to conventional evaporation technique.

1.1.10. Tanning and Leather Industry

In a typical tannery, several unit operations are involved, namely, soaking, liming, pickling, deliming-bating, tanning, dyeing, etc. Each of these steps is extensively chemical consuming and produces large amount of wastewater containing appreciable amount of organic and inorganic materials causing high

COD, BOD, TDS, TS, conductivity etc. This leads to water and soil pollution. With increasing the awareness of environmental conservation, government policy is now becoming stricter everyday and proper treatment of industrial wastewater (specially tannery effluents) has become an important issue. In this regard, membrane based separation processes are found to be very much effective [35]. An integrated process including nanofiltration (NF) followed by reverse osmosis (RO) was developed by Das et al. [36] for the treatment of soaking effluent. Application of UF and RO for treatment of degreasing effluent is also reported [37]. Use of UF and NF for the treatment of effluent from liming and deliming-bating steps is available [38-39].

1.2. Advantage of Membrane Separation Processes

Membrane based separation processes have several advantages over conventional separation processes such as, (i) physical separation process; (ii) no chemicals are required; (iii) less energy requirement; (iv) treatment of heat sensitive materials e.g., fruit juice; (v) mild operating condition (operable under ambient temperature); (vi) almost no damage to the species during processing; (vii) no phase change; (viii) less capital and operating cost; (ix) easy to scale up and (x) requirement of low manpower.

1.3. Limitations of Membrane Separation Processes

Despite considerable progress of the membrane processes, the most serious operational constraint in its industrial application is the flux decline during the operation due to concentration polarization and irreversible membrane fouling.

1.3.1. Concentration Polarization

Concentration polarization refers to the build-up of high concentration of rejected solute within the boundary layer adjacent to the membrane surface [40].The accumulation of solutes at the membrane surface adversely affects the membrane performance. The basic principle of membrane separation leads to an accumulation of the retained solutes and a depletion of the permeating components in the boundary layer adjacent to the membrane. Hence, a concentration gradient develops within the boundary layer. The polarization of the solutes leads to a decrease in the available driving force of the preferentially permeating solutes across the membrane and an increase for the retained solutes. This reduces the overall efficiency of separation.

Consequences of concentration polarization are as follows:

- Increase in solute concentration on the membrane surface leads to an increase in the osmotic pressure of the solution, thereby reducing the effective driving force for the solvent flow [41].
- Formation of cake or gel type layer on the membrane surface that offers an additional resistance to the permeate flow, in series to that of hydraulic membrane resistance [42, 43].
- Changes in physico-chemical properties of solution with solute concentration such as viscosity, diffusivity and density within the boundary layer which adversely affects the permeation rate [44, 45, 46].
- Membrane permeability may be reduced to a larger extent, due to the partial or complete pore blocking [47, 48].

All these factors lead to a decrease in system performance i.e., throughput of the system.

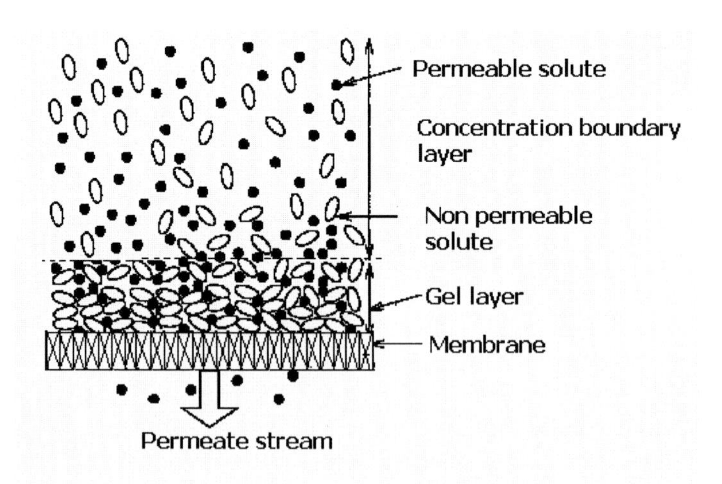

Schematic of Concentration polarization

1.3.2. Irreversible Membrane Fouling

The concentration polarization can facilitate irreversible membrane fouling by altering interaction among solvent, solute and membrane. Membrane pores may be completely or partially blocked by the solute particles. There may also be solute adsorption in the pore mouth or inside the pores leading to pore clogging. All these factors result in the reduction of membrane permeability. This loss of permeability can not be recovered even after thorough washing and thereby causing irreversible fouling.

Membrane fouling can occur by deposition of particles over the membrane surface or within the membrane pore. There are four different kinds of blocking models that are typically used to describe fouling such as complete blocking, intermediate blocking, standard blocking, and cake filtration.

Complete blocking assumes that every particle that reaches the membrane surface completely seals the entrance of the membrane pores and prevents flow. Moreover, a particle never settles on another particle that has previously deposited on the membrane surface. Intermediate blocking is similar to complete blocking. The intermediate blocking model is less restrictive because it assumes that a portion of the particles blocks the pores and the rest accumulate on top of other previously deposited particles. This means that not every particle that arrives to the membrane surface blocks a membrane pore. This model examines the probability of a particle to block a membrane pore. Standard blocking assumes that particles accumulate over the pore walls leading to a decrease in the pore volume. Therefore, the cross section of the membrane pore decreases. As particles are deposited, the pores become constricted and the permeability of the membrane decreases. Cake filtration occurs when particles accumulate on membrane surface in a porous cake of increasing thickness that offers extra resistance to solvent flow. It assumes that each particle is located on others already arrived who blocked some of the membrane pores. The resistance is mainly composed of two parts: the membrane and the cake layer.

1.4. Approaches to Improve Membrane Performance

Concentration polarization and fouling are the major bottleneck, limiting the large-scale industrial implementation of membrane technology. Concentration polarization and membrane fouling largely contribute to the cost of the membrane separation process. Fouling reduces the service life of the membrane, leading to frequent cleaning or replacement of the expensive membrane. Minimization of concentration polarization and membrane fouling is essential to make the membrane processes economically competitive. Concentration polarization cannot be avoided completely in any membrane processes; only its effects can be minimized. To overcome concentration polarization, various techniques have been investigated. Following are the most common: (i) modification of the membrane material, (ii) change of hydrodynamic conditions in the flow channel, e.g, vortex mixing, tube insert, pulsatile flow, gas sparging, reverse flow, corrugated membrane surface (iii) change of physico-chemical environment of the solution (iv) application of external body force, i.e., dc electric field, magnetic field etc. Each of these is briefly discussed as follows.

1.4.1. Surface Modification

A major drawback in the application of ultrafiltration membranes in bio-separations is the protein fouling that can limit the permeate flux during operation. Protein molecules readily adsorb on the membrane surfaces and pore walls leading to the formation of a secondary barrier that decreases permeate flux and changes solute selectivity. It is evident that apart from concentration polarization, membrane-protein interactions as well as protein-protein interactions result in a severe irreversible loss in membrane performance. It is widely accepted that the antifouling characteristics of the hydrophilic membrane are better than those of the hydrophobic one. Hence, modification of the membrane from hydrophobicity to hydrophilicity is very important. Numerous membrane surface modification methods have been investigated. These include chemical treatment, plasma treatment, ion beam irradiation, physical adsorption of modifiers (e.g., surfactants, block copolymers), and grafting etc.

1.4.1.1. Chemical Treatment and Physical Coating

Membrane performance can be improved by modifying its surface using chemical treatment. Surface modification by chemical treatment using hydrophylicity enhancing agent [49], proteic acids [50] are reported in literature. Dimov et al. [50] have reported that permeate flux can be enhanced by chemical modification of microfiltration membrane using a ternary mixture of ethanol-water- inorganic acids. Mukherjee et al. [51] have reported that chemical treatment of polyamide RO membrane with solution of hydrofluoric acid, fluosilicic acid and isopropyl alcohol increases both flux and rejection. Membrane performance can also be enhanced by physical coating of the membrane surface.

Physical coating of polyamide membrane with a specific type polyether-polyamide block co-polymer results higher permeate flux than the uncoated membrane [52].

1.4.1.2. Plasma Treatment

Among the various surface-modification techniques, low temperature plasma treatment is found to be effective to prepare high performance membranes [53, 54] such as ultrafiltration (UF), nanofiltration (NF), reverse osmosis (RO) and gas separation membranes. For the design and development of surface-modified polymer membranes, the capability of plasma to change the physical and chemical properties of polymeric surfaces without affecting the bulk properties of the base material is advantageous. This technique enhances the membrane surface hydrophilicity. Also, permeability of plasma treated membranes is often higher than that of untreated membranes [55, 56]. Plasma processes are broadly classified into plasma treatment and plasma polymerization depending on the type of gases, which are used in plasma. Plasma treatment is non-polymer-forming plasma process; it can be chemically reactive or non-reactive. Plasma polymerization is polymer-forming plasma process; it can lead to form polymeric solid films deposited onto the substrate membrane surfaces.

Plasma treatment is generally carried out using gases like argon, hydrogen, carbon dioxide, helium, oxygen, ammonia etc., [56- 58]. Membranes with rough surface are more susceptible to fouling than smooth surface. Plasma treatment results to surface oxidation and formation of hydrophilic surface leading to smoother surface.

Several investigations on modifications of polymer surfaces by plasma treatment are reported in literature. Hallab et al. [59] explained how polymeric materials demonstrated significant increase in adhesion strength with increased surface roughness. Hoek et al. [60] predicted lower amount of repulsive-interaction energy barrier between a colloidal particle and rough membrane surface, where the reduction in energy barrier is strongly correlated with the magnitude of surface roughness. Aerts et al. [61] showed that how plasma treatment changes the surface energy that plays a major role in membrane fouling and cleaning of a fouled membrane surface. For argon or oxygen plasma, functional groups like peroxide, hydroxyl, carboxylic acid, ketone/aldehyde and ester groups are incorporated on to the membrane surface [62-64]. There are some reports on the use of CO_2 plasma to alter the characteristics of membrane surface [63, 64]. Wavhal et al.[65] have used CO_2 plasma for the modification of polyethersulfone (PES) microfiltration membranes. They have found that treated membranes become more hydrophilic in nature, and there are no changes in the hydrophilicity upon aging, even after storage in air for six months.

1.4.1.3. Ion Beam Irradiation

The effect of ion beam irradiation has a positive impact on the membrane performance [66, 67]. Ion beam irradiation is the bombardment of a substance with energetic ions. As the ions penetrate the membrane, they lose energy to the membrane polymer resulting in bond breaking, cross-linking, and formation of volatile molecules. This alters the microstructure of the polymer and morphology of the surface resulting in a reduction in surface roughness. This reduction in the membrane roughness significantly enhances its performance because smooth membrane surfaces are less prone to fouling than the rough surfaces.

1.4.1.4. Grafting Polymers

To reduce the fouling resistance, membrane surfaces are modified by grafting polymers [11]. One of the widely used graft polymerization technique for membrane surface modification is ultraviolet (UV)-assisted photochemical grafting [68]. Plasma-initiated grafting is another method used for membrane surface modification [69-71]. In this method, the membrane is treated by plasma (e.g., argon plasma), and then, post-plasma treatment, grafting of a hydrophilic polymer is done from the vapor phase. For example, to prepare hydrophilic nanofiltration membranes, low temperature argon plasma treatment and subsequent grafting with acrylic acid has been studied. Redox initiation grafting also has been used for membrane surface modification [72]. The modification leads to increase hydrophilicity, which decreases fouling tendency, while enhancing the permeability and selectivity properties of the membrane. Thermal grafting [73] is also used for surface modification, in which the membrane surface is pretreated to add surface-bound vinyl groups that serve as the anchoring sites for polymer chains that are grown from the surface by thermally activated, free-radical graft polymerization. Surface-initiated atom transfer radical polymerization (ATRP) is also used to modify regenerated cellulose UF membranes. ATRP grafting has been used only recently for the surface modification of gas permeation, microfiltration and pervaporation membranes [74].

1.4.2. Change of Hydrodynamic Conditions

By modification of membrane surface, interaction between solute and membrane can be made favorable to minimize fouling. But, once the cake or gel-type layer is formed, only alternative left is to change the system hydrodynamics so that mass transfer can be improved. Hydrodynamics in the flow channel can be altered either by steady state technique or by imposing instability to the flow. In the steady state technique, high cross flow velocity or stirring can be used.

1.4.2.1. Turbulent Flow

An increase in cross flow velocity in the flow channel is the simplest way to reduce the concentration polarization and fouling on the membrane surface. With increase in cross flow velocity, membrane surface concentration decreases by the forced convection due to increased turbulence and therefore diffusional flux from the membrane increases as the concentration gradient from the surface to the bulk increases leading to the enhancement of permeate flux [40]. For the estimation of mass transfer coefficients, the most common approach for modeling turbulence in membrane separation processes is to use empirical correlations [75]. However, these correlations are system specific and valid in certain ranges of the operating conditions, thereby limiting their applicability. Suction through the porous membrane had a significant effect on the mass transfer coefficient. Prabhakar et al.[76] developed a mathematical model for quantification of concentration polarization phenomena during osmotic pressure controlled ultrafiltration under turbulent flow conditions. Minnikanti et al.[77] have proposed a Sherwood number correlation for the prediction of mass transfer coefficient and permeate flux with suction for turbulent flow condition in a cross flow system. High turbulence may create axial pressure drop which may in turn decreases the transmembrane pressure drop and increases the pumping overhead [78].

1.4.2.2. Unsteady Flows and Induction of Instabilities

Introduction of instabilities in the flow channel results in an enhancement of mass transfer was first noticed by Thomas in 1973 [79]. Hydrodynamic instabilities can be caused by (i) turbulent promoter, (ii) Gas sparging, (iii) secondary flow, and (iv) pulsatile flow.

1.4.2.2.1. Turbulence Promoter

A variety of turbulence promoters have been investigated. The insertion of rods, wire ring, glass beads, kenics mixer, doughnut disk baffles, and moving balls in the flow channel have been identified to minimize fouling [80-83]. The function of turbulence promoter is to create local turbulence in the flow channel by increasing velocity and wall shear rate. Using this technique permeate flux is found to be augmented by several times. Introduction of fluidized particles [84] or intermittent jets [85] in the flow chamber, screw threaded flow promoter [86,87], static mixers [88], natural convection instabilities [89], etc. have also been studied to induce instability in the flow channel. Permeate flux is found to be enhanced significantly.

1.4.2.2.2. Gas Sparging

Gas sparging, the injecting gas bubbles into the feed, has recently been identified as an effective technique to enhance the performance of ultrafiltration and microfiltration membrane [90-95]. The addition of air to the liquid stream increases both the turbulence at the surface of the membrane and the superficial cross flow velocity within the system, suppressing boundary layer formation, leading to an enhancement of the productivity of filtration process. Using this technique, Cui et al. [90] have reported a 250% improvement in flux compared to conventional cross-flow operation for the ultrafiltration of dyed dextran solution. Lee et al. [95] have also reported the use of air slugs to improve the cross-flow filtration of bacterial suspensions.

1.4.2.3. Secondary Flow

Application of dynamic filtration using the rotating cylindrical membrane is found to be useful due to excellent fluid mixing, high shear rate. In rotating membrane, Taylor vortices are formed in the annular portion due to centrifugal flow instability which is useful to control the fouling. Using this technique, significant improvement in permeate flux and solute retention are observed by several investigators [96-98]. High energy consumption for rotating the device, problems in maintenance and replacement of the membrane, scale formation are the major drawback of such rotating device. To overcome these problems, Dean vortices are suggested by several researchers [99,100]. Dean vortices are formed by forcing the fluids in a curved channel at a modified Reynolds number (Dean number). These Dean vortices improve the mass transfer coefficient by reducing concentration polarization leading to an enhancement of permeate flux. Several reports on the use of Dean vortices are available in the literature [101-105].

1.4.2.4. Pulsatile Flow

Application of pulsed flow is found to be effective to reduce concentration polarization and fouling [106]. Pulsed flow may be generated by vibration of porous plate above the membrane surface, pump vibration or ultrasound. Bauser et al. [107] have used periodic sequence of pumping pulses keeping mean flow constant during microfiltration of whey and observed improvement of flux by about 38%. Use of pulsatile flow in RO of sucrose solution [108] and UF and MF of whey protein [109], whole blood flow [110] and protein solutions [111-114] are reported. The entire mass transfer boundary value problem for oscillatory newtonian laminar flow was solved by Illias and Govind [115] and the results were compared with the experimental data of Kennedy et al. [108]. In addition to flow pulsation, transmembrane pressure pulsations were also employed to improve flux [114]. Combination of pulsatile flow with baffles were also investigated [110,111] and were found to improve flux significantly.

1.4.3. Change of Physico-chemical Environment of the Solution

The performance of an ultrafiltration membrane particularly in the separation of charged solutes such as protein, polysaccharide, etc., is strongly influenced by solute-solute and solute-membrane interaction [116]. These interactions strongly depend on the physico-chemical environment of the solution i.e., pH and ionic strength of the solution [117,118]. The extent of fouling and concentration polarization on the membrane surface can be controlled by manipulating pH and ionic strength of the solution. Change in ionic strength or pH of the solution can alter the flux and retention of the same feed solution. The influence of these parameters i.e., pH and ionic strength on fractionation of similar molecular weight proteins has been studied by many authors [119,120]. Surface charges of both the solutes and membrane affect the interaction mechanism which is mainly coulumbic and hydrophobic in nature. Flow through membrane pore under the influence of membrane-solute interaction is investigated by several authors [121-124]. Solute-membrane interaction results the adsorption of solute on the pore wall and clogging of pores by solute aggregates which adversely affects the membrane permeability and retention of solutes [125]. It has also been identified that solute-solute interactions play a major role in the variation of diffusivity and osmotic pressure, which governs the transport of the solvent through the membrane. Solute-solute interactions may be quantified in terms of interaction potential which is mainly composed of Liftshiz-van der waals, polar acid base and electrostatic. Much of the work in this area has been focused on protein-protein interaction. Protein-protein interaction model is based on DLVO theory [126] which provides adequate information for proteins in dilute aqueous electrolyte solution, where the interactions are manly due to long-ranged electrostatic forces.

Osmotic pressure and diffusivity may be estimated using these potential which in term may be used for the prediction of permeate flux.

1.4.4. External Field

Application of external body forces such as acoustic, magnetic, thermal and electric field are found to be quite effective to reduce the extent of fouling and concentration polarization resulting in an enhancement of permeate flux [127,128]. Use of d.c. electric field is the most widely explored area in the application of external field to improve the membrane performance. The electric field imposes an electrophoretic effect on charged solutes dragging them away from the membrane surface. Thus concentration polarization and fouling are reduced leading to an increase in permeate flux. Application of d.c. electric field has been successfully investigated with various systems. These include kaolin suspensions, emulsions of oil in water, bentonite clay, china clay, anastase, BSA, gelatin, xanthane biopolymer, etc., [129-136]. A number of excellent publications on electro-ultra filtration are available in literature. In an early study, a mathematical model based on film theory for cross flow electric field assisted microfiltration under gel polarization domain for kaolin clay suspension and oil-in water emulsion was presented by Henry et al. [129]. Yukawa et al. [133] have studied cross-flow electro ultrafiltration in a tubular module for gelatin solution. Radovich et al. [137] have developed a mathematical model to predict the steady state electro-ultrafiltration flux for processing BSA solution using Amicon XM-50 membrane. Jagannadh and Muralidhara [138] have reviewed, in detail, electro kinetic method to control membrane fouling. Mameri et al. [139] have shown the impact of electric field on fouling caused by BSA during cross flow ultrafiltration of BSA. Huotari et al. [140] have investigated the effect of electric field on flux in cross-flow electro ultrafiltration of oily wastewater. Hofmann et al. [141] have shown the effect of electric field to the improvement of biopolymer recovery and permeate flux in dead-end pressure electro filtration. They have also investigated the effect of pH, ionic strength and applied pressure on the filtration kinetics of a dead-end electro filtration of polysaccharide xanthan suspension. In order to minimize energy consumption, several researchers have used pulsed electric field instead of constant field [142,143,109,110]. Bowen et al. [142] have investigated the use of pulsed electric field to remove the deposited material from stainless steel microfilter and showed that cleaning efficiency increases with increase of applied current. Bowen et al. [143,144] have also utilized pulsed electric field to remove BSA protein deposited on the membrane surface. Microfiltration of dilute suspension of bentonite, anatase and china clay using both constant electric field as well as pulsed electric field are studied by Wakeman and Tarleton [145]. Robinson et al. [146] have shown the potential utility of using pulsed electric field to the enhancement of permeate flux during cross flow ultrafiltration of BSA solution.

Hence, application of external d.c. (direct current) electric field across the membrane is a promising method to reduce concentration polarization and fouling in case of filtration of charged solutes. This proposed book deals with an extremely specialized topic on electric field enhanced membrane separation. This book presents detailed description of the effect of electric field to a number of typical industrial applications, e.g., fruit juice clarification, separation and fractionation of protein solution, micellar enhanced ultrafiltration for dye removal. The results are analyzed in full detail. The relevant principles, technology and the theoretical details are elaborated in the book. Since the topic is an emerging area and most of the work presented has potential of field application, the availability of this book to the researchers will be extremely helpful. Apart from its high academic value, the data presented here will have remarkable impact of scaling up such system in actual industrial scale.

The book presents the sequential discussion of various topics as follows: The application of external d.c. electric field for the improvement of permeate flux in the clarification of fruit juice (synthetic and real juice) is discussed in Chapter 2. A gel polarization model is developed using integral method to quantify the steady state permeate flux. Transient flux decline is also quantified. The effect of pulsed electric field for the enhancement of permeate flux during ultrafiltration of synthetic juice is discussed. Transient flux decline is also studied in presence of pulsed electric field. The methodology for the quantification of deposition thickness on the membrane surface using an optical technique is discussed in this chapter. In case of real mosambi juice, a novel method is proposed to quantify the steady state flux and flux decline in presence of constant as well as pulsed electric field. In Chapter 3, the effect of electric field for the separation of BSA solution and fractionation of protein mixture i.e., BSA and Lysozyme are discussed. An osmotic pressure model using similarity solution is developed to quantify the steady state permeate flux in presence of electric field. The evaluation of membrane surface charge by streaming potential measurement technique is discussed. The role of physico-chemical properties of solution i.e., pH and ionic strength on the permeate flux during ultrafiltration of BSA solution is also discussed. Electric field assisted micellar enhanced ultrafiltration for the removal of methylene blue dye from an aqueous solution is discussed in Chapter 4. A theoretical model is proposed for the prediction of limiting pressure and limiting flux.

REFERENCES

[1] F. Lipnizki, Opportunities and challenges of using ultrafiltration for the concentration of diluted coating materials, *Desalination* 224 (2008) 98-104.

[2] M'nif, S. Bouguecha, B. Hamrouni, M. Dhahbi, Coupling of membrane processes for brackish water, *Desalination* 203 (2007) 331-336.

[3] A. Sonune, R. Ghate, Developments in wastewater treatment methods, *Desalination* 167 (2004) 55-63.

[4] G.A. Pervov, E.V. Dudkin, O.A. Sidorenko, V.V. Antipov, S,A. Khakhenov and R.I. Makarov, RO and NF membrane systems for drinking water production and their maintenance techniques, *Desalination* 132 (2002) 315-321.

[5] G.M. Walker, L.R. Weatherley, Biodegradation and biosorption of acid anthaquinone dye, *Environ. Pollut.* 108 (2000) 219-223.

[6] V. Golob, A. Vinder, M. Simonic, Efficiency of coagulation/flocculation method for the treatment of dyebath effluents, *Dye Pigments* 67 (2005) 93-97.

[7] A. Roessler, D. Crettenand, O. Dossenbach, W. Marte, P. Rys, Direct electrochemical reduction of indigo, Electrochim. Acta 47 (2002) 1989-1995.

[8] S. Wang, Z.H. Zhu, A. Coomes, F. Haghsersht, G.Q. Lu, The physical and surface charaterictics of activated carbons and the adsorption of methylene blue from waste water, *J. Colloid Interface Sci.* 284 (2005) 440-446.

[9] S. Chakraborty, B.C. Bag, S. DasGupta, J.K. Basu, S. De, Prediction of permeate flux and permeate concentration in nanofiltration of dye solution, *Sep. Puri. Tecnol.* 35 (2004) 141-152.

[10] K.M. Pastagia, S. Chakraborty, S. DasGupta, J.K. Basu, S. De, Prediction of permeate flux and concentration of two-component dye mixture in batch nanofiltration, *J. Membr. Sci.* 218 (2003) 195-210.

[11] Tang, V. Chen, Nanofiltration of textile wastewater for water reuse, *Desalination* 143 (2002) 11-20.

[12] Koyuncu, D. Topacik, E. Yuksel, Reuse of reactive dye house waste water by nanofiltration: process water quality and economic implication. *Sep. Purif. Technol.* 36 (2004) 77-86.

[13] R. Atra, G. Vatai, E. B. Molnar, A. Balint, Investigation of ultra- and nanofiltration for utilization of whey protein and lactose, *J. Food Eng.* 67 (2005) 325-332.

[14] G. Brans, C.G.P.H. Schroen, R.G.M. van der Sman, R.M. Boom, Membrane fractionation of milk: state of the art and challenges, *J. Membr. Sci.* 243 (2004) 263–272.

[15] D.N. Lee, R.L. Merson, Prefiltration of cottage cheese whey to reduce fouling of ultrafiltration membrane, *J. Food Sci.* 41 (1976) 403-410.

[16] H.C. van der Horst, J.M.K. Timmer, T. Robbertsen, J. Leenders, Use of nanofiltration for concentration and demineralization in the dairy industry: Model for mass transport, *J. Membr. Sci.* 104 (1995) 205–218.

[17] N.S. Krishna Kumar, M.K. Yea, M. Cheryan, Ultrafiltration of soy protein concentrate: performance and modelling of spiral and tubular polymeric modules, *J. Membr. Sci.* 244 (2004) 235-242.

[18] N.S. Krishna Kumar, M.K. Yea, M. Cheryan, Soy protein concentrates by ultrafiltration, *J. Food Sci.* 68 (2003) 2278-2283.

[19] Girard, L. R. Fukumoto, Membrane processing of fruit juices and beverages: a review. *Critical Reviews in Food Science and Nutrition* 40(2) (2000) 91-157.

[20] Pepper, A.C.J. Orchard, & A.J. Merry, Concentration of tomato juice & other fruit juice by reverse osmosis. *Desalination* 57 (1985) 157-166.

[21] S. Todisco, P. Tallarico and E. Drioli, Modelling and analysis of the effects of ultrafiltration on the quality of freshly squeezed orange juice, *Italian Food and Beverage Technology* 12 (1998) 3-8.

[22] M.L. Wu, R.R. Zall and W.C. Tzeng, Microfiltration and ultrafiltration comparison of apple juices clarification, *J. Food Sci.* 55 (1990) 1162-1163.

[23] A. Cassano, C. Conidi, R. Timpone, M. D'Avella, E. Drioli, A membrane-based process for the clarification and the concentration of the cactus pear juice, *J. Food Eng.* 80 (2007) 914-921.

[24] R. Ghosh, Z.F. Cui, Fractionation of BSA and lysozyme using ultrafiltration: Effect of pH and membrane pretreatment, *J. Membr. Sci.* 139 (1998) 17-28.

[25] R. Ghosh, Z.F. Cui, Purification of lysozyme using ultrafiltration, *Biotechnol. Bioeng.* 68(2) (2000) 191-203.

[26] K. Narasaiah, G.P. Aggarwal, Transmission analysis in ultrafiltration of ternary protein mixture through a hydrophilic membrane, *J. Membr. Sci.* 287 (2007) 9-18.

[27] S. Sakesna, A.L. Zydney, Effect of solution pH and ionic strength on the separation of albumin and immunoglobulin (IgG) by selective filtration, *Biotechnol. Bioeng.* 43 (1994) 960-968.

[28] A. Maartens, E.P. Jacobs, P. Swart, UF of pulp and paper effluent: membrane fouling-prevention and cleaning, *J. Membr. Sci.* 209 (2002) 81-92.

[29] G. Liu, Y. Liu, J. Ni, H. Shi, Y. Qian, Treatability of kraft spent liquor by microfiltration and ultrafiltration, *Desalination* 160 (2004) 131-141.

[30] A.S. Jonsson, R. Wimmerstedt, The application of membrane technology in the pulp and paper industries, *Desalination* 53 (1985) 181-196.

[31] G. Tragardh, V. Gekas, Membrane technology in the sugar industry, *Desalination* 69 (1988) 9-17.

[32] M. Hamachi, B.B. Gupta, R.B. Aim, Ultra filtration: a means for decolorization of cane sugar solution, *Sep. Purif. Technol.* 30 (2003) 229-239.

[33] S. K. Karode, T. Courtois, B.B. Gupta, Ultrafiltration of raw Indian sugar solution using polymeric and mineral membrane, *Sep. Sci. Technol.* 35(15) (2000) 2473-2483.

[34] S.S. Madaeni, S. Zereshki, Reverse osmosis alternative: Energy implication for sugar industry, *Chemical Engineering and processing* 47 (2008) 1075-1080.

[35] A. Cassano, R. Molinari, M. Romano and E. Drioli, Treatment of aqueous effluents of the leather industry by membrane processes A review. *J. Membr. Sci.* 181 (2001) 111-126.

[36] C. Das, S. DasGupta and S. De, Treatment of soaking effluent from tannery using membrane separation processes, *Desalination* 216 (2007) 160-173.

[37] A. Cassano, A. Criscuoli, E. Drioli and R. Molinari, Clean operations in the tanning industry: aqueous degreasing coupled to ultrafiltration: experimental and theoretical analysis. *Clean Prod. Processes* 1 (4) (1999) 257-263.

[38] C. Das, S. De and S. DasGupta, Treatment of liming effluent from tannery using membrane separation processes. *Sep. Sci. Technol.*, 42 (2007) 517-539.

[39] M.T. Ahmed, S. Taha, T. Chaabane, D. Akretche, R. Maachi and G. Dorange, Nanofiltration process applied to the tannery solutions. *Desalination* 200 (2006) 419-420.

[40] W.F. Blatt, A. Dravid, A.S. Michaels, L. Nelson, Solute polarization and cake formation in membrane ultrafiltration: causes, consequences and control techniques, in: J.E. Flinn (Ed.), *Membrane Science and Technology,* Plenum Press, New York, 1970.

[41] S. De, S. Bhattacharjee, A. Sharma, P.K. Bhattacharya, Generalized integral and similarity solution of the concentration profiles for osmotic pressure controlled ultrafiltration, *J. Membr. Sci.* 130 (1997) 99-121.

[42] S. De, P.K. Bhattacharya, Modelling of ultrafiltration process for a two component aqueous solution of low and high (gel forming) molecular weight solutes, *J. Membr. Sci.* 136 (1997) 57-69.

[43] S.K. Karode, A method for prediction of the gel layer concentration in macromolecule ultrafiltration, *J. Membr. Sci.* 171 (2000) 131-139.

[44] W.N. Gill, D.E. Wiley, C.J.D. Fell, A.G. Fane, Effects of viscosity on concentration polarization in ultrafiltration, *AIChE J.* 34 (1988) 1563-1567

[45] R. Field, P. Aimar, Ideal limiting fluxes in ultrafiltration: comparison of various theoretical relationships, *J. Membr. Sci.* 80 (1993) 107-115.

[46] P. Aimar, V. Sanches, A novel approach to transfer limiting phenomena during ultrafiltration of macromolecules, *Ind. Eng. Chem. Fundam.* 25 (1986) 789-798.

[47] K. Kimura, Y. Hane, Y. Watanabe, G. Amy and N. Ohkuma, Irreversible membrane fouling during ultrafiltration of surface water, *Water Research* 38 (14-15) (2004) 3431-3441.

[48] M.R. Wiesner, S. Chellam, The promise of membrane technologies, *Environ. Sci. Technol.* A 33 (17) (1999) 360-366.

[49] M. Nystrom, P. Jarvinen, Nodification of ultrafiltration membrane with UV irradiation and hydrophilicity agent, *J. Membr. Sci.* 60 (1991) 275-296.

[50] A. Dimov and M.A. Islam, Hydrophilization of polyethylene membrane, *J. Membr. Sci.* 50 (1990) 97-100.

[51] D. Mukherjee, A. Kulkarni, W. N. Gill, Chemical treatment for improved performance of reverse osmosis membranes, *Desalination* 104 (1996) 239-249.

[52] J. S. Louie, I. Pinnau, I. Ciobanu, K. P. Ishida, A. Ng, M. Reinhard, Effects of polyether–polyamide block copolymer coating on performance and fouling of reverse osmosis membranes, *J. Membr. Sci.* 280 (2006) 762-770.

[53] P.W. Kramer, Y.S. Yeh, H. Yasuda, *J. Membr. Sci.* 46 (1989) 1.

[54] Michelle L. Steen, Lynley Hymas, Elizabeth D. Havey, Nathan E. Capps, David G. Castner, Ellen R. Fisher, *J. Membr. Sci.* 188 (2001) 97.

[55] Y.J. Wang, C.H. Chen, M.L. Yeh, G.H. Hsiue, B.C. Yu, *J. Membr. Sci.* 53 (1990) 275.

[56] M. Ulbricht, G. Belfort, Surface modification of ultrafiltration membranes by low temperature plasma II. Graft polymerization onto polyacrylonitrile and polysulfone, *J. Membr. Sci.* 111 (1996) 193-215.

[57] T.D. Tran, S. Mori, M. Suzuki, Plasma modification of polyacrylonitrile ultrafiltration membrane, *Thin Solid Films* 515 (2007) 4148–4152.

[58] S. J. Lue, S.Y. Hsiawa, T.C. Wei, Surface modification of perfluorosulfonic acid membranes with perfluoroheptane (C7F16)/argon plasma, *J. Membr. Sci.* 305 (2007) 226-237.

[59] N.J. Hallab, K.J. Bundy, K. O'Connor, R.L. Moses, J.J. Jacobs, Evaluation of metallic and polymeric biomaterial surface energy and surface roughness characteristics for directed cell adhesion, *Tissue Engineering,* 7 (2001) 55-71.

[60] E.M.V. Hoek, S. Bhattacharjee, M. Elimelech, Effect of membrane surface on colloid-membrane DLVO interactions, *Langmuir* 19 (2003) 4836-4847.

[61] S. Aerts, A. Vanhulsel, A. Buekenhoudt, H. Weyten, S. Kuypers, H. Chen, M. Bryjak, L. E.M. Gevers, I.F.J. Vankelecom, P.A. Jacobs, Plasma-treated PDMS-membranes in solvent resistant nanofiltration: Characterization and study of transport mechanism, *J. Membr. Sci.* 275 (2006) 212-219.

[62] J. Lai, B. Sunderland, J. Xue, S. Yan, W. Zhao, M. Folkard, B.D. Michael, Y. Wang, Study on hydrophilicity of polymer surfaces improved by plasma treatment, *Appl. Surf. Sci.* 252 (2006) 3375-3379.

[63] D.S. Wavhal, E.R.Fisher, Modification of porous poly(ether sulfone) Membranes by Low-Temperature CO2 – Plasma Treatment, *J. Polym. Sci. Part B: Polymer Physics,* 40 (2002) 2473-2488.

[64] L.J. Gerenser, J.M. Grace, G. Apai, P.M. Thompson, Surface chemistry of nitrogen plasma-treated poly (ethylene-2,6-naphthalate): XPS, HREELS and static SIMS analysis, *Surf. Int. anal.* 29 (2000) 12-22.

[65] D.S. Wavhal, E.R. Fisher, Hydrophilic modification of poly ether-sulfone membranes by low temperature plasma –induced graft polmerization, *J. Membr. Sci.* 209 (2002) 255-269.

[66] K. Good, I. Escobar, X. Xu, M. Coleman, M. Ponting, Modification of commercial water treatment membranes by ion beam irradiation, *Desalination* 146 (2002) 259-264.

[67] R. Chennamsetty, I. Escobar, X. Xu, Characterization of commercial water treatment membranes modified via ion beam irradiation, *Desalination* 188 (2006) 203–212.

[68] M. Taniguchi, J.E. Kilduff, G. Belfort, Low fouling synthetic membranes by UV-assisted graft polymerization: monomer selection to mitigate fouling by natural organic matter, *J. Membr. Sci.* 222 (2003) 59-70.

[69] M. Ulbricht, G. Belfort, Surface modification of ultrafiltration membranes by low-temperature plasma. 1. Treatment of polyacrylonitrile, *J. Appl. Polym. Sci.* 56 (1995) 325-343.

[70] H. Chen, G. Belfort, Surface modification of poly (ether sulfone) ultrafiltration membranes by low-temperature plasma-induced graft polymerization, *J. Appl. Polym. Sci.* 72 (1999) 1699-1711.

[71] D.S. Wavhal, E.R. Fisher, Hydrophilic modification of polyethersulfone membranes by low temperature plasma-induced graft polymerization, *J. Membr. Sci.* 209 (2002) 255-269.

[72] S. Belfer, R. Fainshtain, Y. Purinson, J. Gilron, M. Nystrom, M. Manttari, Modification of NF membrane properties by in situ redox initiated graft polymerization with hydrophilic monomers, *J. Membr. Sci.* 239 (2004) 55-64.

[73] R.S. Faibish, Y. Cohen, Fouling and rejection behavior of ceramic and polymer-modified ceramic membranes for ultrafiltration of oil-in-water emulsions and microemulsions, Colloids Surf. *A Phys. Eng. Aspects* 191 (2001) 27-40.

[74] N. Singh, Z. Chen, N. Tomera, S. R. Wickramasinghe, N. Soice, S. M. Husson, Modification of regenerated cellulose ultrafiltration membranes by surface-initiated atom transfer radical polymerization, *J. Membr. Sci.* 311 (2008) 225–234.

[75] V. Gekas, B. Hallstrom, Mass transfer in the membrane concentration polarization layer under turbulent cross-flow. Part I. Critical literature review and adaptation of existing Sherwood correlations to membrane operations, *J. Membr. Sci.* 80 (1987) 153.

[76] R. Prabhakar, S. DasGupta, S. De, Simultaneous prediction of flux and retention for osmotic pressure controlled turbulent cross-flow ultra-filtration, *Sep. Purif. Technol.* 18 (2000) 13.

[77] V.S. Minnikanti, S. DasGupta, S. De, Prediction of mass transfer with suction for turbulent flow in cross-flow ultra-filtration, *J. Membr. Sci.* 4079 (1999) 1.

[78] G. Belfort, R.H Davis, A.L. Zedney, The behavior of suspensions and macromolecular solutions in cross microfiltration, *J. Membr. Sci.* 96 (1994) 1-58.

[79] D.G. Thomas, Forced convection mass transfer in hyperfiltration at high fluxes, *Ind. Eng. Chem. Fundam.* 12 (1973) 396-405.

[80] S. Nijarian, B.J. Bellhouse, Enhanced microfiltration of Bovine Blood Using a Tubular Membrane with Screw-Threaded Insert and Oscillatory flow, *J. Membr. Sci.* 112 (1996) 249-261.

[81] A.R. DaCosta, A.G. Fane, D.E. Wiley, Ultrafiltration of whey protein solution in spacer filled flat channel, *J. Membr. Sci.* 76 (1993) 245-254.

[82] M.R. Mackley, N.E. Sherman, Cake filtration mechanisms in steady and unsteady flows, *J. Membr. Sci.* 77 (1993) 113-121.

[83] A. M. Krsti, W. Hoflinger, A. K. Koris, G.N. Vatai, Energy-saving potential of cross-flow ultrafiltration with inserted static mixer: Application to an oil-in-water emulsion, *Sep. Purif. Technol.*57 (2007) 134-139.

[84] G.M. Rios, H. Rakotoarisoa and B.T. de la Fuente, Basic transport mechanisms of ultrafiltration in presence of fluidized particles, *J. Membr. Sci.,* 34 (1987) 331-343.

[85] G. Arroyo and C. Fonade, Use of intermittent jets to enhance flux in cross flow filtration, *J. Membr. Sci.,* 80 (1993) 117-129.

[86] H.R. Millward, B.J. Bellhouse and G. Walker, Screw thread flow promoters: An experimental study of ultrafiltration and microfiltration performance, *J. Membr. Sci.,* 106 (1995) 269-279.

[87] S. Najarian and B.J. Bellhouse, Enhanced microfiltration of bovine blood using a tubular membrane with a screw threaded insert and oscillatory flow, *J. Membr. Sci.,* 112 (1996) 249-261.

[88] J. Haddink, D. Kloosterboer and S. Bruin, Evaluation of static mixer as convection homogene promoters in ultrafiltration of dairy liquids, *Desalination* 35 (1980) 149-167.

[89] K.H. Youm, A.G. Fane and D.E. Wiley, Effects of natural convection instability on membrane performance in dead-end and cross flow ultrafiltration, *J. Membr. Sci.,* 116 (1996) 229-241.

[90] Z.F. Cui, K.I.T. Wright, Gas-Liquid two phase flow ultrafiltration of BSA and Dextran solution, *J. Membr. Sci.* 90 (1994) 183-189.

[91] Z.F. Cui, K.I.T. Wright, Flux enhancement with gas sparging in Downwards Crossflow Ultrafiltration: Performance and Mechanism, *J. Membr. Sci.* 117 (1996) 109-116.

[92] M. Mercier, C. Fonade, C. L. Delorme, How slug flow can enhance the ultrafiltration flux in mineral tubular membrane, *J. Membr. Sci.* 128 (1997) 103-113.

[93] C. Cabassud, S. Laborie, J.M. Laine, How slug flow can improve ultrafiltration flux in organic hollow fibres, *J. Membr. Sci.* 128 (1997) 93-101.

[94] R. Ghosh, Q. Li, Z. Cui, Fractionation of BSA and Lysozyme using ultrafiltration: Effect of gas sparging, *AIChE J.* 44 (1998) 61-67.

[95] C.K. Lee, W.G. Chang, Y.H. Ju, Air slugs entrapped cross-flow filtration of bacterial suspensions, *Biotechnol. Bioeng.* 41 (1993) 525-530.

[96] K.H. Kroner, V. Nissinen, Dynamic filtration of microbial suspensions using an annular rotating filter, *J. Membr. Sci.* 36 (1985) 85-100.

[97] J. Y. Park, C. K. Choi, J. J. Kimb, A study on dynamic separation of silica slurry using a rotating membrane filter Experiments and filtrate fluxes, *J. Membr. Sci.* 97 (1994) 263-273.
[98] S. Lee, R. M. Lueptow, Control of scale formation in reverse osmosis by membrane rotation, *Desalination* 155 (2003) 131-139.
[99] K.Y. Chung, R. Bates, G. Belfort, Dean vortices with wall flux in a curved channel membrane system. Effect of vortices on permeation fluxes of suspensions in microporous membrane, *J. Membr. Sci.* 81 (1993) 139-150.
[100] M. G. Brewster, K.Y. Chung, G. Belfort, Dean vortices with wall flux in a curved channel membrane system. A new approach to membrane module design, *J. Membr. Sci.* 81 (1993) 127-137.
[101] J. N, Ghogomu, C. Guigui, J.C. Rouch, M.J. Clifton, P. Aptel, Hollow-fibre membrane module design: comparison of different curved geometries with Dean vortices, *J. Membr. Sci.* 181 (2001) 71-80.
[102] J. Kaur, G.P. Agarwal, Studies on protein transmission in thin channel flow module: the role of dean vortices for improving mass transfer, *J. Membr. Sci.* 196 (2002) 1-11.
[103] C. Guigui, P. Manno, P. Moulin, M.J. Clifton, J.C. Rouch, P. Aptel, J.M. Laine, The use of Dean vortices in coiled hollow-fibre ultrafiltration membranes for water and wastewater treatment, *Desalination* 118 (1998) 73-79.
[104] S. Srinivasan, C. Tien, Reverse osmosis in a curved tubular membrane duct, *Desalination* 9 (1971) 127-139.
[105] P. Moulin, J.C. Rouch, C. Serra, M.J. Clifton, P. Aptel, Mass transfer improvement by secondary flows: Dean vortices in coiled tubular membranes, *J. Membr. Sci.* 114 (1996) 235-244.
[106] Y.Y. Wang, J.A. Howell, R.W. Field, D.X. Wu, Simulation of cross flow filtration for baffle tubular channels and pulsatile flow, *J. Membr. Sci.* 95 (1994) 243-258.
[107] H. Bauser, H. Chmiel, N. Stroh, E. Walitza, Control of concentration polarization and fouling of membranes in medical, food and biochemical applications, *J. Membr. Sci.*27 (1986) 195-202.
[108] T. J. Kennedy, R. L. Merson and B. J. McCoy, Improving permeation flux by pulsed reverse osmosis, *Chem. Eng. Sci.,* 29 (1974) 1927-1931.
[109] H. Bauser, H. Chmiel, N. Stroh and E. Walitza, Interfacial effects with microfiltration membranes, *J. Membr. Sci.,* 11 (1982) 321-332.
[110] S. M. Finnigan and J. A. Howell, The effect of pulsed flow on ultrafiltration fluxes in a baffled tubular membrane system, *Desalination* 79 (1990) 181-202.

[111] S. M. Finnigan and J. A. Howell, The effect of pulsatile flow on ultrafiltration fluxs in a baffled tubular membrane system, *Chem. Eng. Res. Des.,* 67 (1989) 278-282.

[112] K.D. Miller, S. Weitzel and V.G.J. Rodgers, Reduction of membrane fouling in the presence of high polarization resistance, *J. Membr. Sci.,* 76 (1993) 77-83.

[113] S. Najarian and B.J. Bellhouse, Effect of liquid pulsation on protein fractionation using ultrafiltration processes, *J. Membr. Sci.,* 114 (1996) 245-253.

[114] V.G.J. Rodgers, R.E. Sparks, Reduction of membrane fouling in the ultrafiltration of binary protein mixtures, *AIChE J.*, 37 (1991) 1517-1528.

[115] S. Illias and R. Govind, Potential applications of pulsed flow for increasing concentration polarization in ultrafiltration, *Sep. Sci. Technol.,* 25 (1990) 1307-1324.

[116] R.A. Curtis, L. Lue, A molecular approach to bioseparations: Protein-protein and protein-salt interactions, *Chem. Eng. Sci.* 61 (2006) 907-923.

[117] R. Van Eijndhoven, S. Saksena, A.L. Zydney,Protein fractionation using electrostatic interaction in membrane filtration, *Biotechnol. Bioeng.* 48 (1995) 406-414.

[118] A. Lucas, M.R. Baudry, F. Michel, B. Chaufer, Role of physico-chemical environment on ultrafiltration of lysozyme with inorganic membrane, Colloid Surf. A: Physicochem. *Eng. Aspect* 136 (1998) 109-122.

[119] D.S. Nakao, H. Osada, H. kurata, T. Tsuru, Separation of proteins by charged ultrafiltration membrane, *Desalination* 70 (1988) 191-205.

[120] M. Balakrishnan, G.P Agarwal, Protein fractionation in a vortex filter. II Separation of simulated mixture, *J. Membr Sci.* 112 (1996) 75-84.

[121] D.M. Malone, J.L. Anderson, Hindered diffusion of particles through small pores, *Chem. Eng. Sci.* 33 (1978) 1429-1440.

[122] B.D. Mitchell, W.M. Deen, Theoretical effects of macromolecular concentration and charge on membrane rejection coefficient, *J. Membr Sci.*19 (1984) 75-100.

[123] E.D. Glandt, Density distribution of hard spherical molecules inside small pore of various shapes, *J. Colloid Interface Sci.* 77 (1980) 512-524.

[124] W.M. Deen, Hindered transport of large molecules in liquid-filled pores, *AIChE J.* 33 (1987) 1409-1425.

[125] N. Lee, R.L Merson, Examination of cottage cheese whey by scanning electon microscopy: Relationship to membrane fouling during ultrafiltration, *J. Dairy Sci.* 58(10) (1975) 1423-1432.

[126] E.J.W. Verwey, J.T.G. Overbeek, *Theory of stability of lyophobic colloids,* Elsevier, Amsterdam, 1948.

[127] A. Strzelewicz, Z. J. Grzywna, Studies on the air membrane separation in the presence of a magnetic field, *J. Membr. Sci.* 294 (2007) 60–67.

[128] S. Murad, The role of magnetic fields on the membrane-based separation of aqueous electrolyte solutions, *Chemical Physics Letters* 417 (2006) 465–470.

[129] J.D. Henry, L.F. Lawler, C.H. A. Kuo, A solid-liquid separation process based on cross flow and electrofiltration, *AIChE J.* 23 (1977) 851-859.

[130] S.P. Moulik, Physical aspects of electro-ultrafiltration, *Current Res.* 5 (1971) 771-776.

[131] R.J. Wakeman, E.S. Tarleton, Experiments using electricity to prevent fouling membrane filtration, *Filtration and Separation May/Jun*e (1986) 174-176.

[132] J.M. Radovich, B. Behnam, Concentration ultrafiltration and diafiltration of albumin with an electric field, *Sep. Sci. Technol,* 18(3) (1983) 223-228.

[133] H. Yukawa, K. Shimura, A. Suda and A. Maniwa, Characteristics of cross flow electro-ultrafiltration for colloidal solution of protein, *J. Chem. Eng. Jpn.* 16(3) (1983) 246-249.

[134] H. Yukawa, K. Shrmura, A. Suda, A. Mamwa, Crossflow electro-ultrafiltration for colloidal solution of protein, *J. Chem. Eng. Jpn.* 16(4) (1983) 305-311.

[135] S. Oussedik, D. Belhocine, H. Grib, H. Lounici, D.L. Piron, N. Mameri, Enhanced ultrafiltration of bovine serum albumin with pulsed electric field and fluidized activated alumina, *Desalination* 127 (2000) 59-68.

[136] T. Weigert, J. Altmann, S. Ripperger, Cross-flow electro-filtration in pilot scale, *J. Membr. Sci.*159 (1999) 253-262.

[137] J.M. Radovich, B. Behnam, C. Mullon, Steady state modeling of electro-ultrafiltration at constant concentration, *Sep. Sci. Technol.* 20 (4) (1985) 315- 329.

[138] S.N. Jagannadh, H.S. Muralidhara, Electrokinetic method to control membrane fouling, *Ind. Eng. Chem. Re*s. 35 (1996) 1133-1140.

[139] N. Mameri, S.M. Oussedik, A. Khelifa, D. Belhocine, H. Gharib, H. Lounici, Electric field applied in the ultrafiltration process, *Desalination* 138 (2001) 291.

[140] H.M. Huotari, I.H. Huisman, G. Tragardh, Electrically enhanced crossflow filtration of oily wastewater using membrane as a cathode, *J. Membr. Sci.* 156 (1999) 49-60.

[141] R. Hofmann, C. Posten, Improvement of dead-end filtration of biopolymer with pressure electrofiltration, *Chem. Eng. Sci.* 58 (2003) 3847-3858.

[142] W.R. Bowen, H.A.M. Subuni, Electrically enhanced membrane filtration at low cross-flow velocities, *Ind. Eng. Res.* 30 (1991) 1573-1579.

[143] W.R. Bowen, H.A.M. Subuni, Pulsed electrokinetic cleaning of cellulose nitrate microfiltration membrane, *Ind. Eng. Res.* 31 (1992) 515-523.

[144] W.R. Bowen, R.S. Kingdon, H.A.M. Subuni, Electrically enhanced separation processes: the basis of in situ intermittent electrolytic membrane cleaning (IIEMC) and in situ electrolytic membrane restoration (IEMR), *J. Membr. Sci.* 40 (1989) 219-229.

[145] R.J. Wakeman, E.S. Tarleton, Membrane fouling prevention in crossflow microfiltration by the use of electric fields, *Chem. Eng. Sci.* 42 (1987) 829-842.

[146] C.W. Robinson, M.H. Siegel, A. Condemine, C. Fee, T.Z. Fahidy, B.R. Glick, Pulsed-electric-field cross flow ultrafiltration of bovine serum albumin, *J. Membr. Sci.* 80 (1993) 209-220.

Chapter 2

Clarification of Citrus Fruit Juice

Abstract

This chapter discusses the effect of externally applied d.c electric field during cross-flow ultrafiltration of synthetic juice (mixture of sucrose and pectin) and mosambi (Citrus Sinensis (L.) Osbeck) fruit juice using 50000 (MWCO) polyerthersulfon membrane. Pectin, completely rejected by the membrane, forms a gel type layer over the membrane surface. Under the application of an external d.c. electric field across the membrane, gel layer formation is restricted leading to an enhancement of permeate flux. During ultrafiltration of synthetic juice, application of d.c. electric field (800 V/m) increases the permeate flux to almost threefold compared to that with zero electric field. A theoretical model based on integral method assuming suitable concentration profile in the boundary layer is developed. The proposed model is used to predict the steady state permeate flux in gel-layer governed electric field assisted ultrafiltration. A gel polarization model is also proposed incorporating continuous as well as pulsed electric field and numerically solved to quantify the flux decline and growth of the gel layer thickness. The gel layer thickness is also measured with high-resolution video microscopy and successfully compared with results from the numerical solution of the model under various operating conditions. Predictions of the model are successfully compared with the experimental results under a wide range of operating conditions.

Application of d.c. electric field during ultrafiltration of mosambi juice has resulted in significant improvement of permeate flux. A steady state gel polarization model has been applied to evaluate the gel layer concentration, effective diffusivity and effective viscosity of the juice within the concentration boundary layer, by optimizing the experimental flux data. Application of pulsed electric field is found to be more beneficial in terms of energy consumption, and equally effective in terms of flux augmentation as compared to constant field.

NOMENCLATURE

A	Constant defined by Eq.(2.25)
A'	Membrane surface area, m^2
$a_{1,2,3}$	Coefficients in Eq.(2.29)
c	Concentration of pectin, kg/m^3
c_o	Pectin concentration at bulk, kg/m^3
c_g	Pectin concentration in gel, kg/m^3
c^*	Dimensionless concentration (c / c_o)
c_g^*	Dimensionless gel concentration (c_g / c_0)
d_p	Equivalent spherical diameter of pectin, m
d_e	Equivalent diameter, m
D	Diffusivity of pectin, m^2/s
D_i	Dielectric constant, dimensionless
E	Electric field, V/m
E_{total}	Total energy consumption, kWh/m^3
E_{pump}	Pump energy consumption, kWh/m^3
$E_{electrical}$	Energy consumption due to external electric field, kWh/m^3
I	Current, amp
h	Half of channel height, m
K	Boltzmann constant (1.38×10^{-23}), J/K
k	Mass transfer coefficient, m/s
L	Channel length, m
M_w	Molecular weight, kg/k.mol
N	Fraction of on time, dimensionless
P	power consumption per unit volume of permeate, watt hr/m^3
Q	permeation rate, m^3/s

$\overline{Pe_w}$	Average peclet number (dimensionless flux)
Pe_e	Peclet number due to electric field (dimensionless flux)
Re	Reynolds number at the bulk condition
Sc	Schmidt number at the bulk condition
T	Absolute temperature, °C
t	Time, s
t_1	On-time, s
t_2	Off-time,s
t_{total}	Total operational time, s
u_e	Electrophoretic mobility, ms^{-1}/Vm^{-1}
v_e	Electrophoretic velocity, m/s
v_w	Permeate flux, m^3/m^2 s
$v_{w, pure}$	Pure water flux, m^3/m^2 s
v_w^{exp}	Experimental permeate flux, m^3/m^2 s
v_w^{cal}	Calculated permeate flux, m^3/m^2 s
u	Axial velocity, m/s
u_0	Average bulk velocity, m/s
v	Velocity component in normal direction, m/s
x	Coordinate from the membrane, m
y	Normal distance, m
y*	Dimensionless normal distance (y/h)
x*	Dimensionless axial distance (x/L)
z	Parameter in Eq. (2.12), $kg/kmol.m^3$

Greek Letters

μ	Viscosity, Pa s
δ_g	Gel layer thickness, m
δ_c	Thickness of concentration boundary layer, m
δ_c^*	Dimensionless concentration boundary layer thickness (δ_c / h)
ζ	Zeta potential of particle, V
ε_o	Permittivity of vacuum (8.854×10^{-12}), $CV^{-1}m^{-1}$
ρ	Bulk density, kg/m^3

ρ_g	Gel density, kg/m^3
ΔP	Transmembrane pressure drop, kPa
ΔP_{pump}	Pump pressure drop, kPa
ρ_g	Gel density, kg/m^3
η_{pump}	Efficiency of pump
$\eta_{d.c.\ power\ supply}$	Efficiency of d.c. power supply

2.1. Importance and Techniques of Clarification of Fruit Juice

India is the second largest producer of fruits in the world, but only 2% of the produced fruits are processed. In India, the production per capita availability is very low (120-130 g/day against 300g recommended) due to wastage caused by inadequate post-harvest facilities. Hence to minimize the post harvest losses due to spoilage and to meet the increasing market demand, there is a need to process the fruit juice with increased shelf-life and to retain the properties of fresh fruit, as well as color, aroma, nutritional value and structural properties. The composition of fruit juice depends on the variety, origin, weather, processing procedure and storage. The major components of fruit juice are sugar (in the form of sucrose, glucose, fructose, etc.), salts, acids, proteins, pectin, starch, minerals, vitamins, flavor and aroma compounds. The organic flavor and aroma compounds contain alcohols, esters, aldehydes etc., which are highly volatile in nature. The raw fruit juice obtained after extraction is very viscous, turbid, rich in color and contains significant amount of colloidal materials (100-1000 ppm) which are stabilized in suspension by pectin, starch etc., [1]. Hence, raw juice needs to be clarified prior to its commercial use. Clarification of juice involves the removal of haze–forming components such as suspended solids, colloidal particles, pectins, proteins etc. For clarification of fruit juice, traditional methods involve several labor-intensive and time consuming batch operations, e.g., use of enzyme (pectinase), fining agent (gelatine, bentonite etc.), filter aids (diatomaceous earth, kieselguhr etc.) etc. However, the use of these additives (fining agent, filter aids) give bitter taste in the juice. Moreover, the solids obtained after filtration contains enzyme, fining agent and filter aids. This can not be reused and causes pollution problems due to their disposal [2]. For concentrating clarified juice, evaporation is the most common technique but it causes the loss of aroma and flavor compounds and requires high energy consumption [3].

In this regard, membrane based separation processes have gradually become a cost effective alternative to conventional fining and clarification technique. These processes are athermal and involve no phase change or addition of chemicals. Apart from it, other advantages of these processes are low energy consumption, less processing time, production of additive free high quality juice with natural fresh taste.

The major problem associated with membrane based clarification of fruit juice is the flux decline due to concentration polarization and membrane fouling during the operation. The objective of ultrafiltration of fruit juice is to retain high molecular weight pectin and its derivative (which have a tendency to form gel on the membrane surface) and to allow low molecular weight solute like sucrose, acid, salt etc., to permeate through the membrane. In fruit juice, typical concentration of pectic substances is up to 1.0 % [4]. Pectins are acidic polysaccharides that originate in the cell membrane structure in plants which is composed of a rahmno-galacturonan backbone in which 1-4 linked α-D-galacturonan chain are interrupted and bent at the intervals by the insertion of 1-2 linked α-L-rahmnopyranosyl residue. One of the important properties of pectin is that it forms gel in presence of sugar and acid. Pectin gel is nothing but network of cross-linked polymer molecules in a liquid medium [5]. The strength and characteristics of pectin gel depend on degree of methoxylation, presence of sugar, average molecular weight, ionic strength, pH, temperature and charge density etc. Based on degree of esterification, pectins are classified into high methoxyl-pectin (>50% esterified) and low methoxyl-pectin (<50% esterified). High methoxyl-pectins form gel in presence of sugar and acid (pH<3.6) by the formation of hydrogen-bonding and hydrophobic interactions. Gelation of low methoxyl-pectin occurs in presence of divalent cations, such as calcium ion, which binds the pectin molecules and acts as a bridge between two carboxyl groups of pectin [5]. Apart from gel forming characteristics, other important properties of pectin is that it carries a negative charge and its charge density depends on pH and degree of methoxylation. The variation of zeta potential with pectin concentration at different solution pH and glucose contents is discussed by Sulaiman et al. [6]. Gelation of pectin is largely affected by the charge distribution in rahmno-galacturonan backbone.

Gel forming pectin molecules are found to behave as significant organic foulants of membranes [7]. Therefore, depectinization of juice is a common pretreatment method before membrane clarification. In order to reduce the pectin content, an enzymatic treatment of the raw juice is usually carried out with enzyme such as pectinase. Pectinase hydrolyses pectin resulting in a flocculation of pectin-protein complex. This enzyme treatment reduces the amount of pectin in the juice and makes the filtration process easier by lowering its viscosity. During ultrafiltration of fruit juice, the left over pectin even after enzyme treatment is sufficient to cause a formation of gel type layer over the membrane surface which offers extra resistance to the permeate flow in series with the hydraulic resistance of the membrane [8,9] and thereby drastically reduces the permeate flux. Hence, improvement of permeate flux during fruit juice clarification is an interesting challenge.

2.2. APPLICATION OF EXTERNAL FIELD

Application of various external fields such as acoustic, magnetic, and d.c. electric field for the reduction of concentration polarization and membrane fouling have been discussed in Chapter 1. Among the various external fields, use of d.c. electric field is explored here to improve the membrane performance. Several studies have been focused to reduce concentration polarization and membrane fouling but very few reports are available in literature regarding the use of external electric field to the enhancement of permeate flux. In electro-ultrafiltration, the concentration polarization is reduced by applying an external d.c. electric field across the membrane. A number of excellent publications on electro-ultra filtration are available in literature (see section 1.4.4). The major problems associated with the use of continuous electric field during ultrafiltration include changes in feed properties due to electrode reaction [10-13], requirement of high energy, rise of temperature, etc. The method can not be effectively used for feeds of high electrical conductivity and heat sensitivity. These problems can be circumvented by the use of pulsed electric field instead of a continuous one. A few references of the use of pulsed electric fields are available in literature (see section 1.4.4). During electro-ultrafiltration, electrolysis may take place at the electrode. When electric field is applied, the passage of current causes the reaction at the two electrodes and may lead to change in pH near the electrode. Rabie et al. [14] and Vijh [15] have discussed the possible electrode reactions in detail. Most possible electrode reactions are as follows:

Reduction at the cathode:

$$2H_2O + 2e^- \rightarrow H_2 + 2OH^- \tag{2.1}$$

$$M_c^{n+} + ne^- \rightarrow M_c \tag{2.2}$$

Oxidation at the anode:

$$M_a \rightarrow M_a^{n+} + ne^- \tag{2.3}$$

$$6H_2O \rightarrow O_2 + 4H_3O^+ + 4e^- \tag{2.4}$$

where, M_c and M_a represent the cathode and anode material, respectively. Reaction (2.1) leads to the evolution of hydrogen gas and OH^- ions near the cathode. Reaction (2.2) takes place if the cathode metal is positioned below hydrogen in the electrochemical series. Since the porous cathode plate is placed below the membrane, the produced OH^- ions are washed out with the permeate stream resulting an increase in pH of the permeate. Reaction (2.3) can be avoided by using inert material such as platinum, silver, gold, etc. and reaction (2.4) causes the evolution of oxygen gas and H_3O^+ that leads to decrease of pH near the anode.

2.2.1. Application of Electric Field during Fruit Juice Clarification

Pectin being negatively charged, application of external d.c. electric field during cross-flow ultrafiltration of fruit juice appears to be promising. Application of an appropriate external electric field across the membrane significantly reduces the gel-type layer formation by the charged pectin molecules, thereby increasing the permeate flux (and possible reduction of membrane fouling). This observation has encouraged us to investigate the effect of electric field on the permeate flux in the clarification of fruit juice.

Several models have been attempted to simulate the ultrafiltration performance of fruit juice. The most conventional models are gel-polarization model [16], osmotic pressure model [17], boundary layer model [18], resistance in series model etc. Shen et al. [19] have developed a similarity solution model and Probstein et al. [20] have proposed an integral method to model gel layer controlled ultrafiltration. Permeation of sucrose through a gel forming material (PVA) is studied and modeled by De et al. [16]. In contrast, a suitable mathematical model for electric field enhanced ultrafiltration of gel forming particle is rare in literature.

In this chapter, the potential of an externally applied electric field for the reduction of pectin gel type layer formation over the membrane surface has been explored. A cross-flow ultrafiltration in presence of electric field has been carried out under a wide range of operating conditions for the clarification of synthetic juice (a mixture of pectin and sucrose) and mosambi (*Citrus Sinensis (L.) Osbeck*) juice. Effects of operating conditions, e.g. electric field, feed concentration, pressure and cross flow velocity on the permeate flux are investigated in detail. The integral method of solution under the frame work of boundary layer analysis of Probstein et al. [20] is extended for the prediction of steady state permeate flux in gel controlled ultrafiltration of synthetic juice in presence of an external d.c. electric field. The gel type layer thickness during ultrafiltration of a synthetic juice solution of pectin and sucrose mixture is measured using high-resolution video microscopy. A theoretical model for the growth of the gel-type layer over the membrane surface under various operating conditions is proposed, numerically solved and successfully compared with optically measured values of gel thickness. It is appeared from the earlier discussion that use of a pulsed electric field may be able to address some of the inherent limitations and shortcomings associated with the application of a continuous field. There has been no report in the literature about the use of pulsed electric field for fruit juice clarification. The present work also investigates to study the effects of pulsed electric field, cross flow velocity and feed concentration on the permeate flux during clarification of synthetic and mosambi juice. A theoretical model based on resistance-in-series model for prediction of permeate flux and gel layer thickness over the membrane surface is proposed in presence of pulsed electric field. Mosambi juice is a complex mixture of several components, the viscosity and diffusivity values are difficult to obtain. The gel concentration of the actual juice is also different from that of pure pectin. Moreover, the variation of viscosity and diffusivity within concentration polarization layer near the membrane surface is significant due to sharp variation in the concentration. For the difficulty in estimating the concentration dependence of viscosity and diffusivity of the real juice, an optimization technique has been developed to evaluate the effective values of these parameters. The value of mass transfer coefficient required for the above mentioned optimization has been deduced from the Sherwood number relationship derived in integral solution method for synthetic juice. Knowing these parameters, transient flux decline is quantified by a resistance-in-series model during ultrafiltration of mosambi juice in presence of continuous as well as pulsed electric field.

2.3. Membrane Experiment

2.3.1. Cross-flow Electro-ultrafiltration Cell

The experiments are conducted in continuous mode of operation. From the feed tank, feed solution is pumped and allowed to flow tangentially over the membrane surface through a thin channel of 37 cm in length, 3.6 cm in width and 6.5 mm in height. The anode is platinum coated titanium sheet (length 33.5 cm, width 3.4 cm, thickness 1.0 mm) obtained from Ti Anode Fabricators, Chennai (India), and mounted parallel to the flow position just above the ultrafiltration channel. External electric field from a regulated DC power supply is applied across the membrane surface. The retentate is recycled to the feed tank. The flow rate is measured by a rotameter in the retentate line. Pressure inside the electro ultrafiltration cell is maintained by operating the bypass valve and measured by a pressure gauge. Permeate is collected from bottom of the cell. The effective filtration area is 133.2 cm^2. The details of schematic of the experimental set up are shown in Figure 2.1 and 2.2.

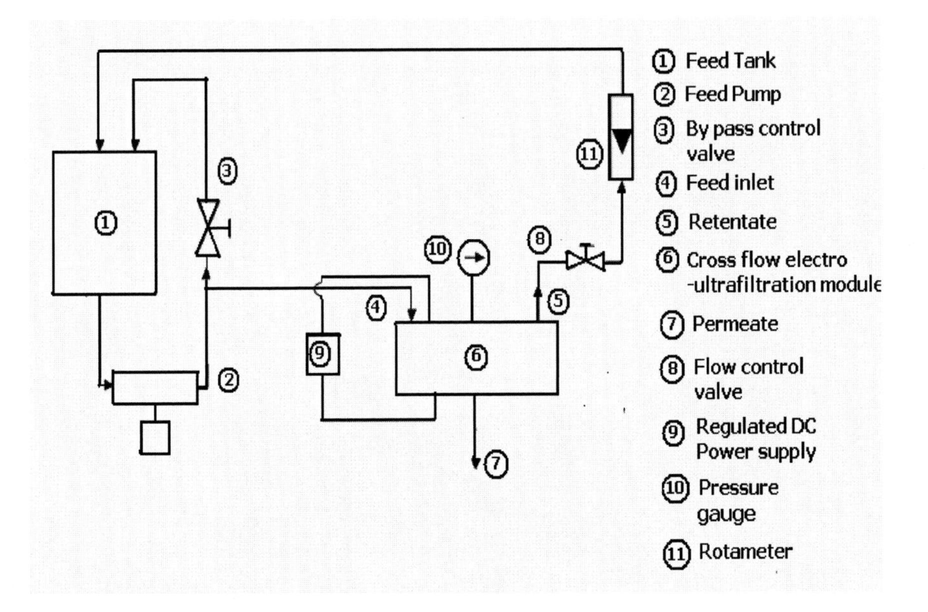

Figure 2.1. Schematic diagram of electro-ultrafiltration system.

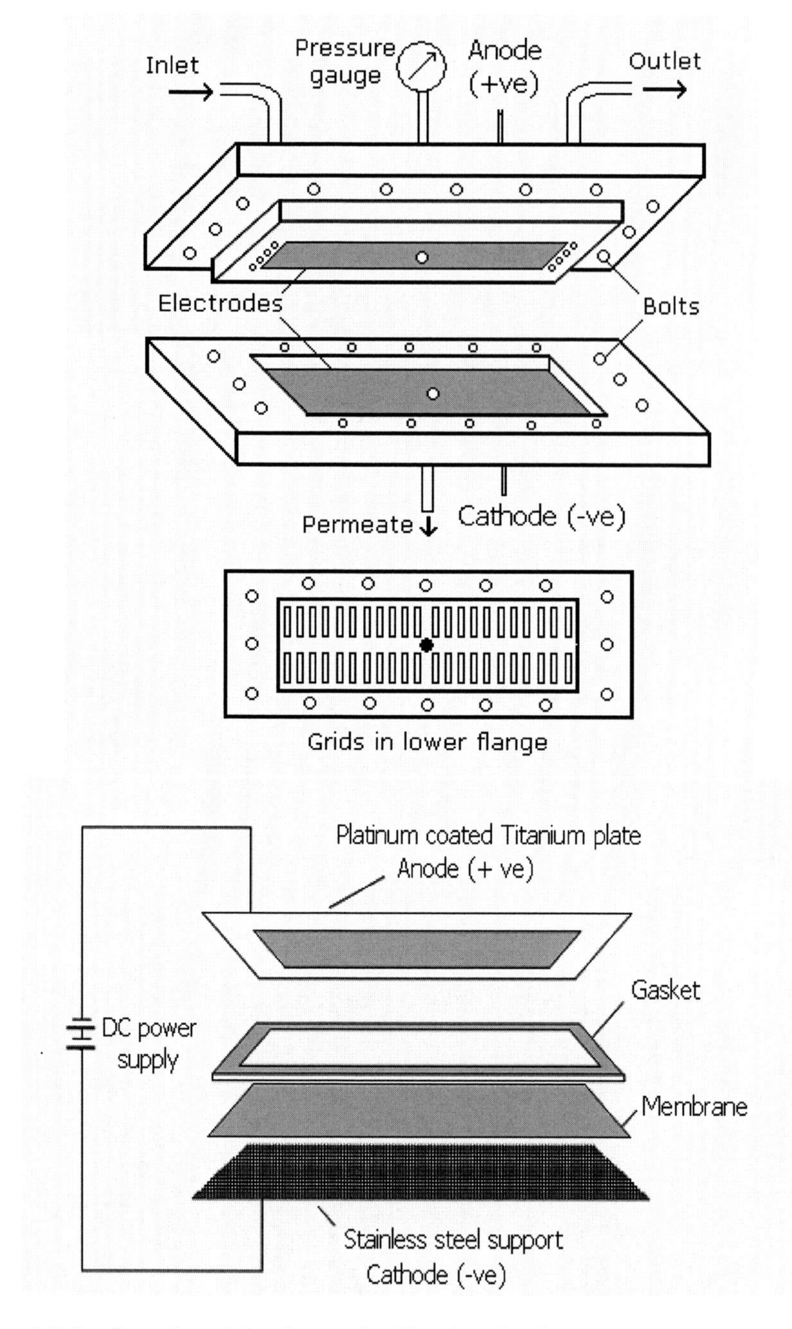

Figure 2.2.Configuration of the electro-ultrafiltration chamber.

2.3.2. Procedure for Conducting Experiments

2.3.2.1. Materials

Mosambi (*Citrus sinensis (L.) Osbeck*) fruits were collected from local market. In industries, membranes having a wide range of pore sizes, with molecular weight cut-off (MWCO) 18 kDa to 0.2 μm are generally used for clarification of fruit juice [21]. Since the average molecular weight of pectin is approximately 65 kDa, flat sheet polyethersulfone membrane of MWCO 50 kDa is chosen for electric field assisted clarification of synthetic juice as well as mosambi juice. Membranes obtained from M/s, Permionics Membranes Pvt. Ltd., Boroda, Gujarat, (India), have been used for the experiments.

2.3.2.2. Operating Conditions

Electro-ultrafiltration experiments are designed to observe the effects of variations in operating conditions (transmembrane pressure, cross-flow velocity, electric field, feed concentration and pulse ratio). The electrical conductivity of mosambi fruit juice (4.0 m S/cm) is significantly higher compared to that of the synthetic juice (0.17 m S/cm). Due to the high electrical conductivity of mosambi juice, use of higher electric field results in significant water hydrolysis at the electrodes leading to evolution of oxygen (O_2) and hydronium ions (H_3O^+) at the anode and hydrogen (H_2) and hydroxyl ions (OH^-) at the cathode. This may change the properties of the feed as well as the permeate and may result in the rise of temperature. To avoid electrolytic reactions, maximum operating electric field has been restricted to 500 V/m.

2.3.2.3. Preparation of Feed

Three different mixtures of pectin and sucrose are prepared in double distilled water (Pectin: 1 kg/m^3, Sucrose: 14 Brix; Pectin: 3 kg/m^3, Sucrose: 12 Brix; Pectin: 5 kg/m^3, Sucrose: 10 Brix) and used for the experiments. Mosambi juice was extracted by a manually operated 'screw type juice extractor'. Pectinase from *Aspergillus nigar* with activity 3.5-7.0 units/mg protein (Lowry) (SIGMA-ALDRICH (USA) was used to treat the mosambi juice at an enzyme concentration of 0.01% v/v at 44°C (in a bath at constant temperature) for 120 minutes. After the treatment, the suspension was heated to 90°C for 5 minutes in a water bath to inactivate the remaining enzyme in the sample. The decanted, depectinized juice was used for electro-ultrafiltration.

Table 2.1. Operating conditions used in experiment

Juice	Variable	Operating conditions
Synthetic juice	Feed	Pectin: 1 kg/m^3; Sucrose: 14 Brix Pectin: 3 kg/m^3; Sucrose: 12 Brix Pectin: 5 kg/m^3; Sucrose: 10 Brix
	Transmembrane pressure (kPa)	220, 360, 500, 635
	Cross flow velocity (m/s)	0.09, 0.12, 0.15, 0.18
	Electric field (V/m)	0, 300, 600, 800,1000, 1200,1600, 2000
	Pulse ratio (off time always set to 1 sec).	1:1, 2:1, 3:1, 4:1, constant electric field
Real juice	Mosambi juice	
	Transmembrane pressure (kPa)	220, 360, 500, 635
	Cross flow velocity (m/s)	0.09, 0.12, 0.15, 0.18
	Electric field (V/m)	0, 200, 300,400, 500
	Pulse ratio (off time always set to 1 sec).	1:1, 2:1, 3:1, 4:1, constant electric field

2.3.2.4. Conduction of Experiments

The membrane is compacted at a pressure of 690 kPa (higher than the highest operating pressure used in this study) for 3 hours using distilled water. Pure water fluxes at different operating pressures are measured next and plotted against pressure difference. The membrane permeability is obtained from the slope of this curve as $1x10^{-10}$ m/Pa s for 50 kDa. During ultrafiltration experiment, permeate samples are collected at different time. The duration of each experiment is 30-40 minutes. All experiments are conducted at $30\pm2^{\circ}C$. After each experiment, the membranes are cleaned using the following procedure:

After completion of an experiment membrane was washed thoroughly for 45 minutes by recirculating distilled water at room temperature $30\pm2^{\circ}C$, followed by 45 minutes of static acid (pH 3.0) washing using HCl, 30 minutes of washing using distilled water, 30 minutes of alkaline (pH 10) washing and finally membrane was washed with distilled water. After such thorough washing, water flux was measured with distilled water. This procedure is allowed for recovery of the pure initial water flux within 95%. All the steps are carried out at room temperature i.e., $30\pm2^{\circ}C$. The membranes are also cleaned in situ in some of the experiments in the same way as mentioned before. In both the cases results are almost same. It has been found that the original permeability of the membrane is completely recovered in most cases. Application of external d.c. electric field has a significant effect on membrane fouling. In presence of electric field the deposition over the membrane surface is less and therefore less time is needed for its cleaning compared to without electric field.

2.3.2.5. Analysis of the Feed and Permeate

The concentrations of pectin and sucrose in the synthetic juice as well as in the permeate and retentate are determined using a Genesys2 Spectrophotometer. For pectin concentration, a wavelength of 230 nm is used and distilled water is taken as a blank. For pectin-sucrose mixture, the pectin free solution containing same amount of sucrose is used as blank. The sucrose contents in the sample are assessed with a refractometer (Thermospectronic). The raw mosambi juice and clarified juice samples were analyzed for color, clarity, total soluble solid, titrable acidity, pH, viscosity, pectin content, and conductivity. Color was measured by absorbance at 420 nm and clarity by transmittance at 660 nm using Genesys2 Spectrophotometer (Thermospectronic, USA). The total soluble solid was measured by refractometer, (Thermospectronic, USA). The acidity of the sample was determined by titration against 0.1M NaOH and expressed as % of anhydrous citric acid. The viscosity was measured by using an Ostwald viscometer. Zeta potential is obtained by measuring electrophoretic mobility and then by using Henry equation. The electrophoretic mobility of the particle is measured by Laser Doppler Velocimetry. The conductivity is measured by an autoranging conductivity meter, Toshniwal Instrument (India). Alcohol insoluble solid (AIS) was used as a measure for pectin content in the juice. AIS values were measured by mixing 20 g juice with 300 ml 80% alcohol solution and simmering the mixture for 30 minutes. The filtered residue was washed with 80% alcohol solution. The residue was dried at 100°C for 2 hours and is expressed in weight percentage [22]. Pectin content of the juice was about 0.38 times of AIS value as obtained from a calibration curve. The zeta potential of mosambi juice at its natural pH was measured by a zeta meter (Zetasizer, Nano-ZS, Malvern Instruments, UK). Physico-chemical properties of synthetic juice and mosambi juice are presented in Table 2.2 and 2.3.

The particle size of the mosambi juice are determined by Malvern Zetasizer (Nano ZS, ZEN3600, Malvern, England). It measures the particle size using the technique of dynamic light scattering (DLS). The suspended particles are in brownian motion. DLS measures the brownian motion of the particle to calculate the size of the particle.

Table 2.2. Physico-chemical properties of synthetic juice

Pectin + Sucrose mixture	pH	Conductivity (mS/cm)	Viscosity x 10^3 (Pa s)	Density x 10^{-3} (kg/m^3)	Zeta potential (mV)	Diffusivity x 10^{11}(m^2/s)
(1 kg/m^3 + 14 Brix)	3.5	57.0	1.83	1.06	-19.4	5.11
(3 kg/m^3 + 12 Brix)	3.2	127.0	2.72	1.05	-20.5	3.44
(5 kg/m^3 + 10 Brix)	3.16	170.0	3.10	1.04	-20.1	3.02

Table 2.3.1. Physico-chemical properties of mosambi juice

Juice	Pectin (kg/m^3)	TSS (^{0}Brix)	pH	Density $\times 10^{-3}$ (kg/m^3)	Viscosity $\times 10^3$ (Pa s)	Conductivity (mS/cm)
Mosambi juice	2.5	8.5	4.10	1.04	2.45	4.0
Mosambi juice (enzyme treated)	1.6	8.5	4.15	1.03	1.05	4.0

Table 2.3.2. Physico-chemical properties of mosambi juice

Juice	Color	Clarity	Acidity as Citric acid (kg/m^3)	Zeta potential (mV)
Mosambi juice	1.33	38.50	7.1	-21.0
Mosambi juice (enzyme treated)	0.96	40.50	5.4	-21.8

In this system Helium-Neon red laser is used to provide the source of light at a wavelength of 638.2 nm and a detector is used to measure the intensity of scattered light at an angle of 173°, called backscattered detection. The rate of intensity fluctuation is used to calculate the size of the particle. Thus DLS gives a measure of intensity particle size distribution shown in Figure 2.3. The average particle size are found to be 1.676 μm, 1.495 μm and 0.145 μm for actual mosambi juice, enzyme treated juice and permeate respectively.

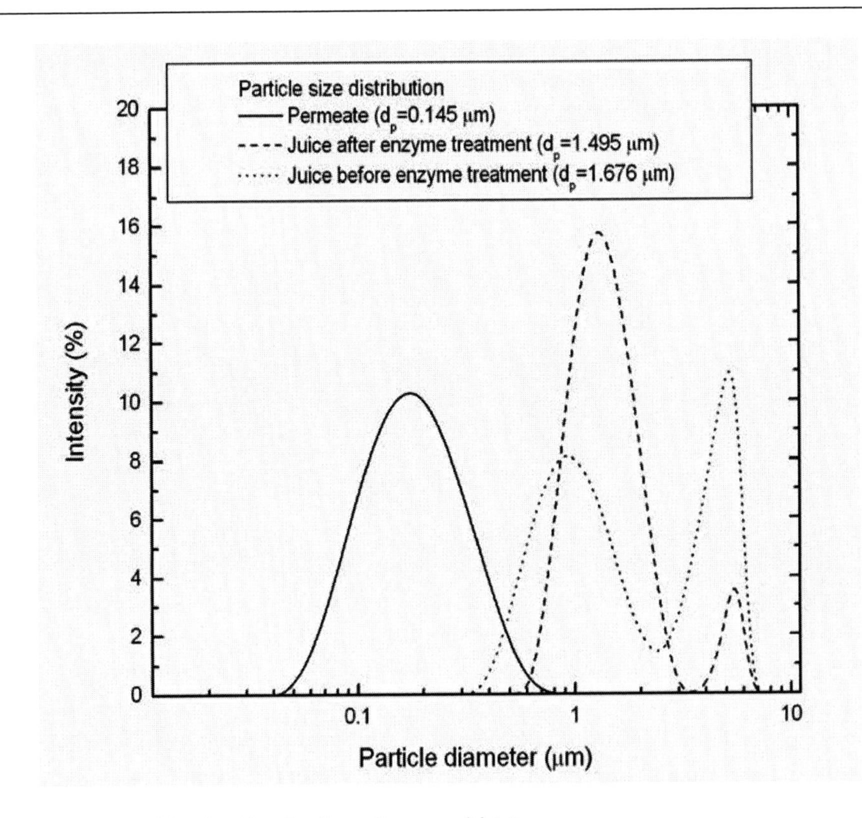

Figure 2.3. Particle size distribution of mosambi juice.

Variation of Zeta Potential of Synthetic Juice (A Mixture of Pectin and Sucrose) with pH

Table 2.4 illustrates the variations of zeta potential with different pH values of pectin sucrose solution, indicating the significant increase of the zeta potential with increasing suspension pH. From the table it is clear that with increasing pH, pectin becomes more negative. Similarly, for a fixed value of solution pH, the pectin becomes more positive with increasing sucrose contents. For an example, at a pectin concentration of 1 kg/m^3, the measured zeta potential decreases from -35 mV to -54 mV with an increase of pH from 3.5 to 9.0; whereas, at pH 5.0, zeta potential increases from -56.2 mV to -39.0 mV with the addition of 14 Brix sucrose to the pectin suspension. The presence of sucrose in pectin suspension causes the electrostatic repulsion between the pectin molecules to decrease and also promotes hydrophobic interaction between the ester methyl groups of pectin and hence facilitates the formation of stable gel [6].

Table 2.4. Values of zeta potential of synthetic juice for various solution pH

Pectin: 1 kg/m^3, Sucrose: 14Brix		Pectin: 3 kg/m^3, Sucrose: 12Brix		Pectin: 5 kg/m^3, Sucrose: 10 Brix	
pH	Zeta potential (mV)	pH	Zeta potential (mV)	pH	Zeta potential (mV)
3.5	-35.0	3.5	-26.6	3.5	-19.5
4.0	-51.0	3.8	-40.6	3.8	-32.8
6.0	-56.2	4.0	-45.0	4.0	-36.1
7.0	-53.7	5.0	-46.5	5.0	-40.0
7.9	-53.5	7.0	-42.8	7.0	-36.5
9.0	-54.0	9.0	-41.6	9.0	-34.0

2.3.2.6. Optical Studies

Immediately after each experimental run, the filtration chamber is opened, carefully removing the top but leaving the lower gasket undisturbed that essentially hold the gel formed along with a layer of the solution. The soft gel over the membrane is partially dried by illuminating the surface with high power lamps to prevent any unwanted motion and loss during transportation for final drying in vacuum desiccators. After two hours of vacuum drying, the deposited membranes are then cut carefully to avoid any beveling or physical depression in the cutting edge. A large number of pieces are then cut from relevant locations. The sample pieces are cut in 4mm x 3mm size pieces. Each sample piece is fixed vertically with its edge held up at the edge of a glass slide. The glass slide with the membrane piece projecting out is then placed under a high-resolution optical microscope (ε POL 400 of Nikon, Japan). Multiple images are obtained of the top of the projected part of the membrane (which essentially is the cross section of the original deposited membrane) and further analyzed to quantify deposition. For each such mounted membrane piece, images are captured at large number of locations and averaged to eliminate local fluctuations. The sampling of the membrane after each experiment is done carefully to encompass the whole area of the membrane by analyzing at many points on the two diagonals of the rectangular membrane piece.

The optical measurements are done in reflection mode of incident light. The membrane sample pieces are viewed under the microscope as well as in a LCD display interfaced with the microscope and a high-resolution digital camera. The captured images are digitized and then analyzed to evaluate the thickness of different layers of the membrane, voids in the substrate material, deposition over the membrane surfaces. A standard grating element having 2000 lines per inch is used to calibrate the magnification of the microscope (500 X) coupled with the camera.

2.4. Prediction of the Permeate Flux and Deposition Thickness

2.4.1. Analysis of Transient Flux

2.4.1.1. Theoretical Aspects

The schematic of a cross flow electro-ultrafiltration system along with development of gel-layer and external concentration boundary layer with and without external d.c. electric field are shown in Figure 2.4 (a and b) respectively. During ultrafiltration, solutes are convected by the transmembrane pressure gradient towards the membrane surface.

The lower molecular weight solutes permeate through the membrane whereas; higher molecular weight solutes are retained, forming a gel-layer over the membrane surface. Since the bulk concentration of higher molecular weight solute is much less than that of the gel layer concentration, a concentration boundary layer forms from the bulk of the solution up to the gel layer. This prompts diffusion of gel forming solutes from the gel layer towards the bulk of the solution due to concentration gradient. The formation of external concentration boundary layer and gel layer over the membrane surface without electric field is shown in Figure 2.4a. In presence of external d.c. electric field (with the top surface maintains opposite polarity as that of the charged particles), the particles move away from the gel layer due to electrophoresis. At steady state, the rate of convective movement of solute particles towards the membrane surface is equal to the rate of migration of solute particles away from the membrane surface due to both back diffusion and electrophoresis (Figure 2.4b).

The formation of external concentration boundary layer and gel layer over the membrane surface without electric field is shown in Figure 2.4a. In presence of external d.c. electric field (with the top surface maintains opposite polarity as that of the charged particles), the particles move away from the gel layer due to electrophoresis. At steady state, the rate of convective movement of solute particles towards the membrane surface is equal to the rate of migration of solute particles away from the membrane surface due to both back diffusion and electrophoresis (Figure 2.4b).

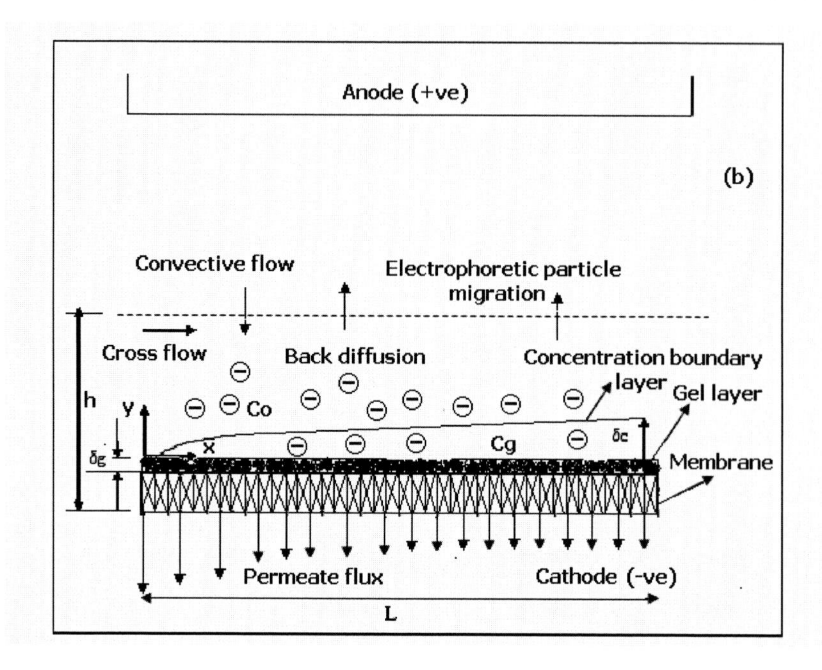

Figure 2.4. Schematic of formation of external concentration boundary layer and gel layer over the membrane surface (a) with out electric field (b) with electric field.

As has been pointed out before, high-methoxylated pectin, as is used herein for experiments, can form gel even in the absence of divalent cations such as Ca^{+2}, if the conditions are acidic and in presence of sugar. Visual observations of the membrane after the experiments confirm the presence of a gel-type layer on the surface. Furthermore, the osmotic pressure contribution of the feed solution is estimated, based on Vant Hoff's equation, and is found to be significantly small compared to the applied transmembrane pressure (less than 1%). Experimentally the permeate flux exhibits almost a pressure independence further indicating that the filtration is carried out under gel-controlled regime. Based on the above considerations, the model equations are developed. In Figure 2.4a, δ_g is the gel layer thickness and δ_c is the thickness of the concentration boundary layer over the gel layer. c_g and c_0 are the gel concentration and bulk concentration of solute respectively. Solute concentration in permeate (c_p) can be neglected by considering very high rejection of solute (by selecting a membrane with low molecular weight cutoff)

The rate of gel layer formation on the membrane surface is given by a solute mass balance as,

$$\rho_g \frac{dL}{dt} = \left(v_w - v_e\right)c + D\frac{dc}{dy} \tag{2.5}$$

The concentration of gel layer is considered to be constant. Therefore, the relevant boundary conditions are

$$c = c_0 \text{ at } y = \delta_g + \delta_c \tag{2.6}$$

$$c = c_g \text{ at } y = \delta_g \tag{2.7}$$

The electrophoretic velocity (v_e) is related with the electrophoretic mobility by the following equation,

$$v_e = E\, u_e \tag{2.8}$$

where, E is the applied electric field. The electrophoretic mobility, u_e can be expressed by Helmholtz-Smoulochowski's equation [23] as

$$u_e = \frac{\varepsilon_0 D_i \xi}{\mu} \tag{2.9}$$

where, ε_o is permittivity of vacuum, D_i is the dielectric constant and μ is the viscosity. Eq. (2.5) can be solved using the boundary conditions, to obtain the growth of the gel layer thickness as

$$\rho_g \frac{dL}{dt} = \left[\frac{c_g - c_0 \exp\left(\frac{v_w - v_e}{k}\right)}{1 - \exp\left(\frac{v_w - v_e}{k}\right)}\right]\left(v_w - v_e\right) \tag{2.10}$$

where, k is the mass transfer coefficient, defined as $\frac{D}{\delta_c}$ and D is diffusion coefficient. The diffusion coefficient is given by Einstein equation as,

$$D = \frac{KT}{3\pi\mu d_p} \tag{2.11}$$

where, K, T, μ, d_p are the Boltzmann constant, absolute temperature, viscosity of the solution and equivalent spherical diameter of the particle respectively. The average diameter of the gel forming pectin molecule is estimated assuming spherical molecule and using the following relationship [24]

$$M_w = z\, d_p^3 \tag{2.12}$$

where, M_w is the average molecular weight of gel forming material, the value of z taken as $6\text{x}10^{29}$ gm/gm-mole.m^3 [24]. For synthetic juice, the pectin used has a molecular weight range from 30-100 kDa and hence 65 kDa is taken as an average molecular weight. The corresponding value of equivalent spherical diameter of pectin is found to be $4.77\text{x}10^{-9}$ m. The gel concentration of pectin used is 38 kg /m^3 [25]. Hence, permeate flux can also be expressed by a resistance in series formulation including the flow resistance due to the membrane itself and the deposited gel layer. For pure water, the flux can be expressed as

$$v_{w,\,pure} = \frac{\Delta P}{\mu R_m} \tag{2.13}$$

In case of a gel forming macromolecular solution,

$$v_w(t) = \frac{\Delta P}{\left[\mu\left(R_m + R_g(t)\right)\right]} \tag{2.14}$$

where, R_m and R_g are membrane hydraulic resistance and gel layer resistance respectively.

Gel layer formation is similar to that of gel formation in classical filtration. Thus the gel layer resistance (R_g) can be expressed in term of gel layer thickness as,

$$R_g(t) = \alpha(1-\varepsilon_g)\rho_g L(t) \tag{2.15}$$

Where α is specific gel resistance which is obtained from Kozeny-Carman equation as,

$$\alpha = 180 \frac{(1-\varepsilon_g)}{\varepsilon_g^3 d_p^2 \rho_g} \tag{2.16}$$

where, ε_g , ρ_g and d_p are the porosity, density, and diameter of the gel forming particles respectively.

The specific gel layer resistance (α) and the permeate flux are obtained by solving the model equations in an iterative manner following the sequence of steps as outlined below.

1. For known input parameters of operating conditions, membrane parameter and physical properties such as ΔP, c_o, E, R_m, c_g, Di, μ, ρ_g, ξ , ε_o, d_p, a value of specific gel layer resistance (α) is guessed.
2. In case of constant electric field ultrafiltration, applied electric field remains constant throughout the operation. In case of pulsed electric field ultrafiltration, at a particular operational time, t, the applied electric field (either E or 0) is determined and used for subsequent calculations.
3. v_e is calculated using Eqs.(2.8) and Eq.(2.9).
4. V_w^{cal} is calculated by solving the coupled differential and algebraic equations, Eqs. (2.10), (2.14), (2.15) and (2.16) including the effect of pulsed electric field (i.e. constant electric field during on-time and zero electric field during off-time) with the initial condition, L = 0 at t = 0.
5. The time, t, is increased to t +Δt next.
6. Steps 2 to 4 are repeated till t = t_{total}.

7. If $\sum_{i=1}^{N_{total}} \left(\frac{V_w^{exp} - V_w^{cal}}{V_w^{exp}} \right)^2 \leq 0.001$ the program is terminated and transient profiles of permeate flux are obtained. The corresponding value of α is recorded. If not another value of α is guessed in step1 and the process is continued till the given convergence is achieved. It needs to be pointed out that the values of the specific gel resistance,α, cannot be calculated a-priori, and the porosity ε_g , can be a function of the operating conditions (e.g. pressure).

2.4.1.2. Analysis of Transient Flux Decline of Synthetic Juice

Effect of Constant Electric Field

As discussed in earlier, the value of specific gel layer resistance, α is estimated byminimizing the root mean square error associated measured and calculated flux values.

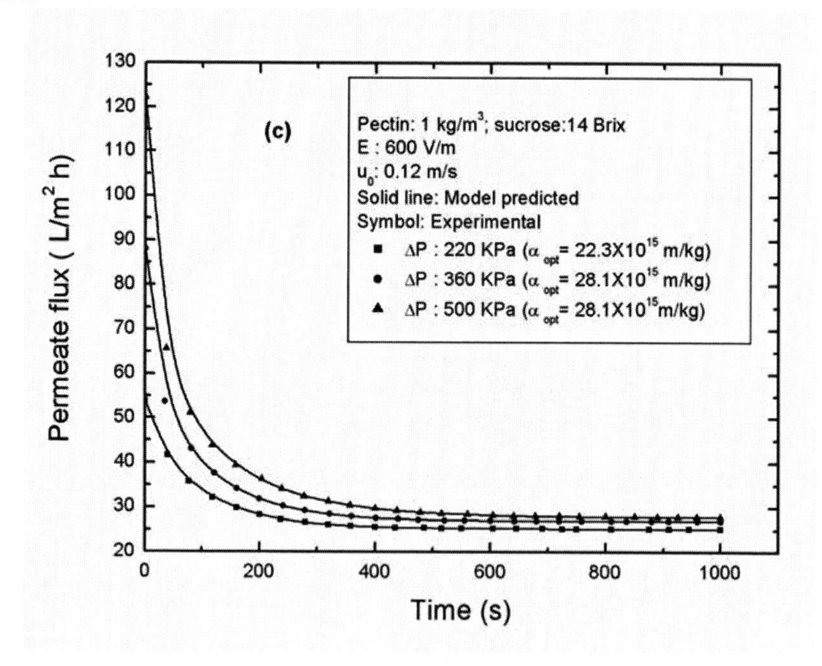

Figure 2.5. Variation of permeate flux with time for synthetic juice (Pectin: 1 kg/m^3 + Sucrose: 14 Brix) (a) E = 800 V/m, u_0 = 0.09 m/s; (b) E = 800 V/m, u_0 = 0.15 m/s; (c) E = 600 V/m, u_0 = 0.12 m/s.

The profile of permeate flux is calculated. It may be observed that the value of α increases with operating pressure but at higher pressure it becomes invariant (Table 2.5). The probable cause for this observation is that the gel layer behaves like a compressible one up to 360 kPa and its compressibility becomes almost invariant for higher pressure. For a constant gel concentration, the specific gel layer resistance, α, varies with operating pressure which is observed in the present study as well as other studies; e.g., Bhattacharjee et al. [26]. The specific gel layer resistance, α is related with gel porosity by Kozeny-Carman equation. Since α typically increases with pressure (resulting in gel compaction), porosity shows a decreasing trend with increase in pressure.

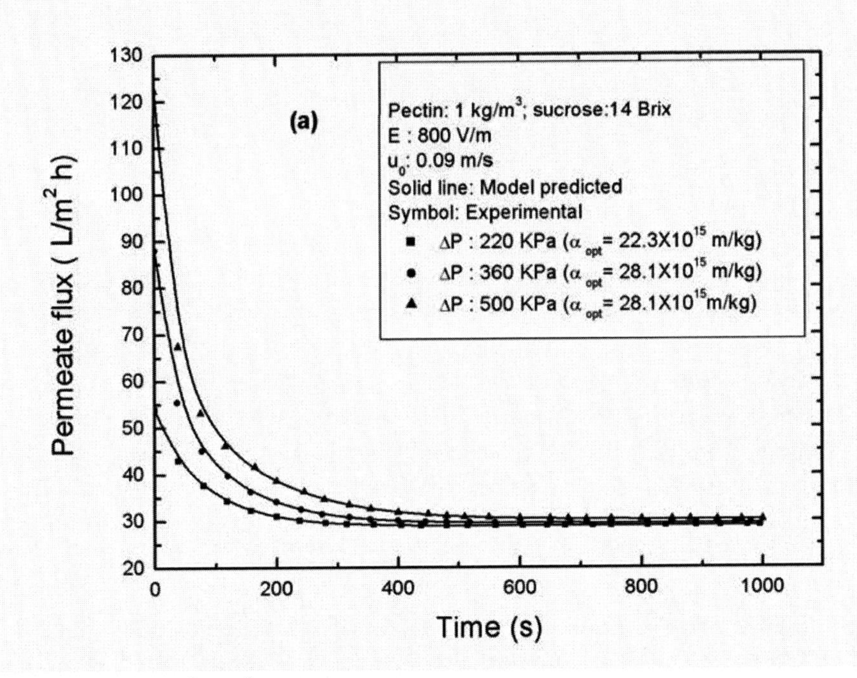

Figure 2.5. (Continued).

Figure 2.5 (a, b and c) to 2.6 (a, b and c) represent the transient profile of permeate flux and gel layer thickness at different operating pressure differences for fixed values of feed concentration, electric field and cross flow velocity. The symbols are the experimental data and solid lines are the model predictions. The flux values are higher for higher operating pressure, increased as expected, due to increased convection.

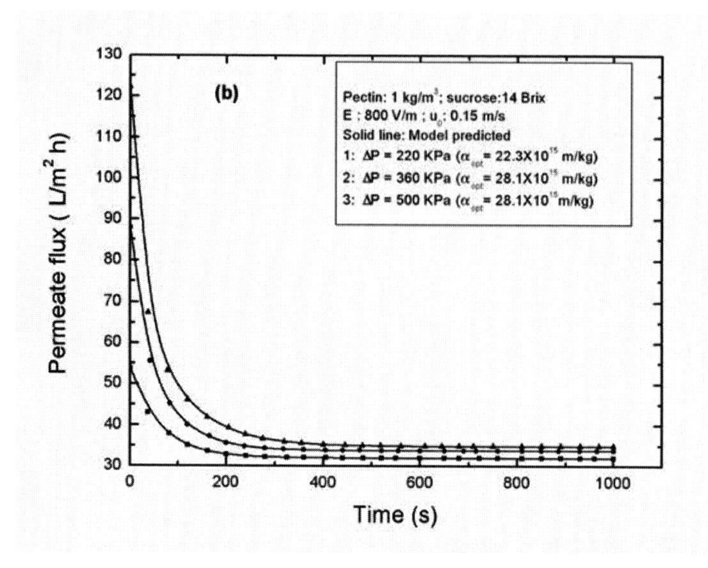

Figure 2.5. (Continued)

The initial rapid decline in flux is due to the building up of concentration polarization and subsequent formation of gel layer over the membrane. It is clear that steady state values in terms of flux are reached within 3 to 4 minutes. From the figures it is clear that gel layer thickness decreases with increase in applied electric field. Applied electric field reduces the gel layer formation by electrophoresis of the pectin molecules in the reverse direction i.e., opposite to the convective flow of filtrate resulting in decrease of gel layer thickness. It is evident from the figure that the thickness (steady state) becomes about half at 800 V/m compare to that of the no electric field case.

Table 2.5. Variation of specific gel layer resistance (α) with operating pressure for synthetic juice

Transmembrane pressure, ΔP (kPa)	Specific gel layer resistance, **α** (m/kg)
220	22.3×10^{15}
360	28.1×10^{15}
500	28.1×10^{15}

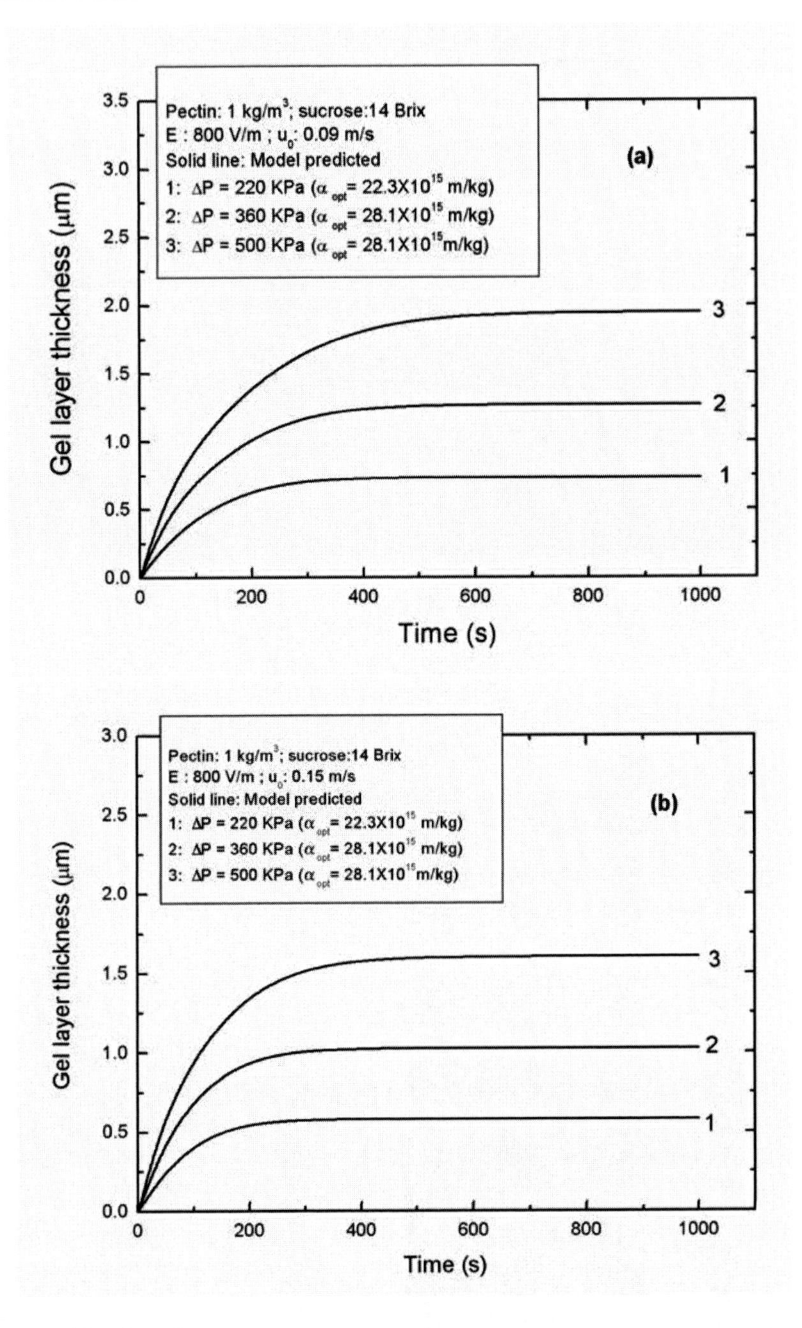

Figure 2.6. (Continued)

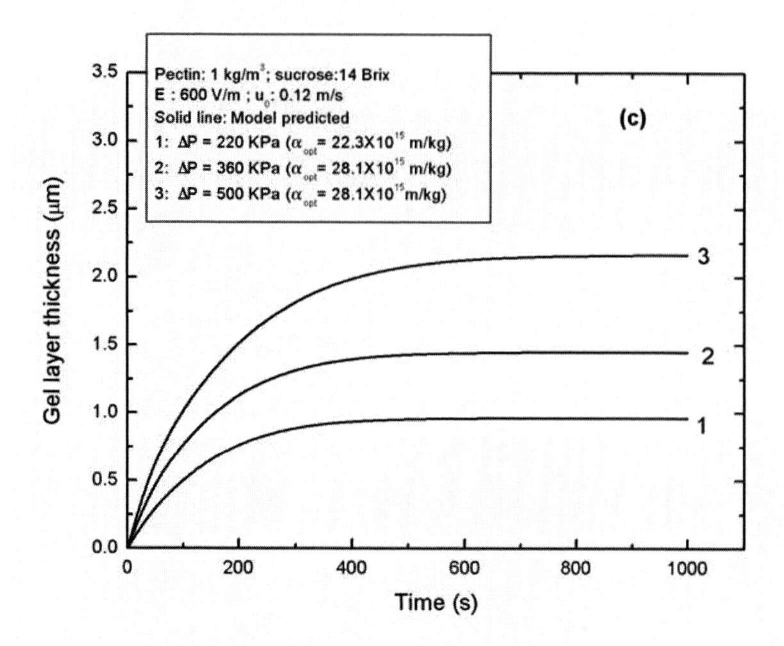

Figure 2.6. Variation of gel layer thickness with time for synthetic juice (Pectin: 1 kg/m^3 + Sucrose: 14 Brix) (a) E = 800 V/m, u_0 = 0.09 m/s; (b) E = 800 V/m, u_0 = 0.15 m/s; (c) E = 600 V/m, u_0 = 0.12 m/s.

Figure 2.7 and 2.8 represent the effect of applied electric field on permeate flux and gel layer thickness. The symbols are experimental data and solid lines are predictions from the model. The sharp decrease observed at the beginning can be attributed to the rapid adsorption of pectin molecules on the membrane and buildup of concentration polarization at the beginning of the experiment. As the time of operation progresses, there is more accumulation of solutes over the membrane surface leading to severe concentration polarization. The smoother and slower flux decline at later stages can be attributed to gel layer formation on the membrane surface. With applying electric field, the pectin molecules tend to move away from the membrane surface due to electrophoresis and gel layer formation is reduced leading to substantial increase of permeate flux. For example, the steady state flux at an electric field of 800 V/m is 18.0 L / m^2 h, showing about 200 % increase over the zero electric field flux of 5.9 L / m^2 h (Figure 2.7). With the same increment of electric field gel layer thickness decreases from 8.88 μm to 2.315 μm (Figure 2.8)

Optical Quantification of Gel Layer Thickness

The images of the deposited membrane after electro-ultrafiltration are analyzed using software, Radical Microcam of Radical Instruments (India) to obtain the thickness of the deposit. The measurements are taken at several positions and then averaged to eliminate local fluctuations. The microscopic observation confirms the decrease in deposition thickness with increase in applied electric field while other parameters remain constant. For example, the deposition decreases from 8.63 microns to 5.52 microns (Figure 2.9) with increase in electric field from 0 V/m to 800 V/m. for a fixed feed concentration (3 kg/m^3+12 Brix), pressure (360 kPa) and cross-flow velocity (0.09 m/s). The gel layer thicknesses for all the ten operating conditions reported in this study are measured optically. However a calibration is required before comparison of optically measure gel layer thickness and those predicted from the numerical solution of the model [27,28]. The comparison between the predicted values of the gel thickness and the experimental measurement is depicted in Table 2.6 which clearly shows the excellent match between these two sets. Table 2.6 also summarizes the effect of different operating parameters. For example the highest gel layer thickness is observed in the case with zero electric field. As expected, it can be seen from the table that gel layer thickness decreases with increasing electric field, increasing cross-flow velocity and decreasing feed concentration. For example, the gel layer thickness (steady state) at an electric field of 800 V/m is 2.069 μm compared to 5.3 μm for no electric field case keeping other operating conditions (pressure, flow rate and feed concentration) unchanged. Gel layer thickness increases with feed concentration. For example, gel layer thickness increases from 1.68 μm to 3.238 μm with increases in feed concentration from 1 kg/m^3 pectin to 5 kg/m^3 pectin while other operating conditions remain unaltered. Again for a fixed value of electric field and pressure, with increases in cross flow velocity from 0.09 m/s to 0.15 m/s, gel layer thickness (steady state) decreases from 2.582 μm to 2.062 μm.

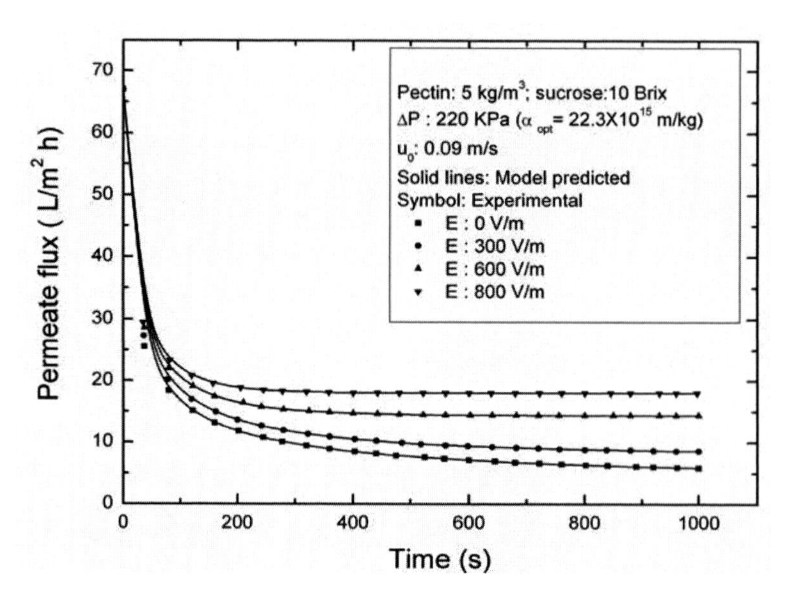

Figure 2.7. Variation of permeate flux with time for different electric fields during clarification of synthetic juice.

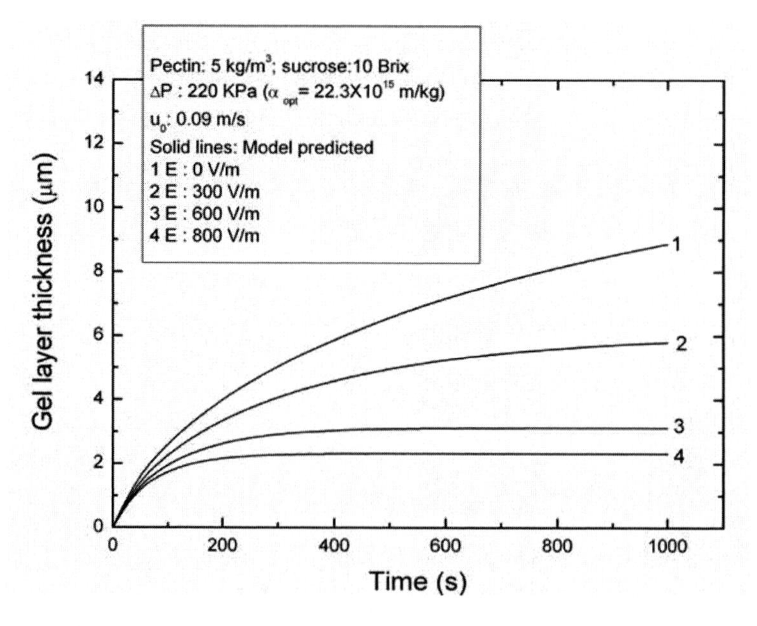

Figure 2.8. Variation of gel layer thickness with time for different electric fields during clarification of synthetic juice.

Table 2.6. Comparison of predicted and experimental gel thickness values

Different parametric conditions	Gel porosity, ε_g	Theoretically obtained gel thickness from flux values (A) (micron)	Modified values of optically measured gel height D = B-C (micron)
(3kg/m^3 +12 Brix) ΔP : 360 kPa u_0: 0.09 m/s E: 300 V/m	0.4775	3.619	*
(3kg/m^3 +12 Brix) ΔP : 360 kPa, u_0: 0.09 m/s E: 600 V/m	0.4775	2.582	*
(.3kg/m^3 +12 Brix) ΔP : 220 kPa, u_0: 0.09 m/s E: 600 V/m	0.5062	2.042	*
(3kg/m^3 +12 Brix) ΔP : 500 kPa, u_0: 0.09 m/s E: 600 V/m	0.4775	3.42	*
(3kg/m^3 +12 Brix) ΔP : 360 kPa, u_0: 0.12 m/s E: 600 V/m	0.4775	2.173	*
(3kg/m^3 +12 Brix) ΔP : 360 kPa, u_0: 0.15 m/s E: 600 V/m	0.4775	2.062	2.073
(5kg/m^3 +10 Brix) ΔP : 360 kPa u_0: 0.09 m/s E: 600 V/m	0.4775	3.238	3.363
(1kg/m^3 +14 Brix) ΔP : 360 kPa, u_0: 0.09 m/s E: 600 V/m	0.4775	1.68	1.573
(3kg/m^3 +12 Brix) ΔP : 360 kPa, u_0: 0.09 m/s E: 0 V/m	0.4775	5.3	5.193
(3kg/m^3 +12 Brix) ΔP : 360 kpa, u_0: 0.09 m/s E: 800 V/m	0.4775	2.069	2.083

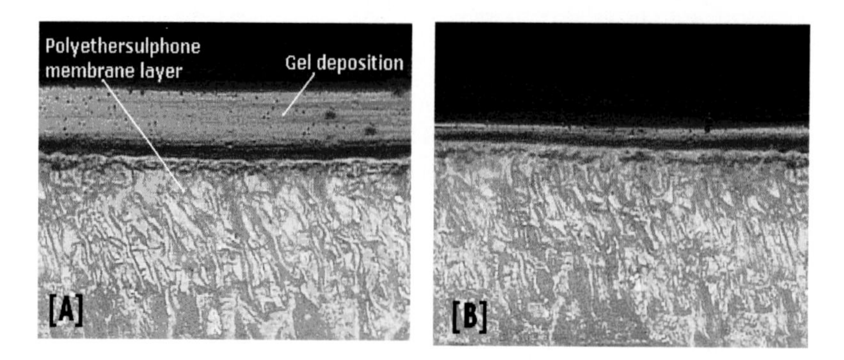

Figure 2.9. Reduction of gel layer thickness with applied d.c. electric field for a synthetic juice (Pectin: 3 kg/m^3, Sucrose: 12 Brix), pressure (360 kPa) and cross flow velocity (0.09 m/s); (A) E = 0 V/m, (B) E = 800 V/m.

Effect of Pulsed Electric Field

The variation of permeate flux with time for different pulse (on-time/off-time ratio) ratio at a transmembrane pressure of 360 kPa and cross flow velocity of 0.12 m/s during clarification of synthetic juice (a mixture of pectin: 3 kg/m^3 and sucrose: 12 Brix) is shown in Figure 2.10

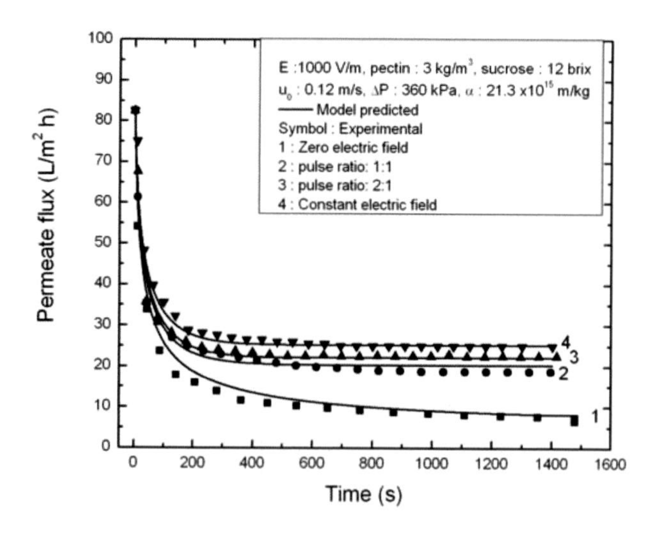

Figure 2.10. Variation of permeate flux with time for different on pulse ratio during clarification of synthetic juice.

2.4.1.3. Mosambi (Citrus Sinensis (L.) Osbeck) Juice

Effect of Constant Electric Field

The variation of permeate flux and gel layer thickness with time for different electric field at a transmembrane pressure of 360 kPa and cross flow velocity of 0.12 m/s during clarification of mosambi juice is shown in Figures 2.11 and 2.12 respectively. The symbols are experimental data and solid lines are model predictions. For a given set of operating conditions, the value of specific gel resistances (α) is obtained by minimizing the root mean square error between the measured and calculated flux values. For mosambi juice, at a transmembrane pressure of 360 kPa, cross flow velocity of 0.12 m/s, and electric field of 400 V/m, the optimum value of α is found to be 2.34 $\times 10^{16}$ m/kg. It is also observed that value of α is invariant with electric field. It is evident from the figure that permeate flux declines rapidly as the gel layer grows and reaches a steady value. The same trend can be observed for all operating conditions. It may be observed from the figure, that at an electric field of 400 V/m, at the end of operation flux declines from 130 L/m^2 h to 11.95 L/m^2 h and gel layer thickness grows up to 3.8 μm whereas, for zero electric field flux declines to 8.67 L/m^2 h and gel layer grows up to 5.4 μm, denoting a sharp rise in steady state permeate flux with electric field application.

The values of specific gel resistances (α), at an operating condition is selected based on the best-fit value i.e. by minimizing the root mean square error associated measured and calculated flux values. The corresponding values of ε_g are calculated from Eq. 2.16. The variation of specific gel resistances (α) and gel porosity (ε_g) with operating pressure are shown in Table 2.7. It may be observed that the value of α increases sharply for lower pressure and gradually for higher pressure values. For example, with increase in pressure from 220 kPa to 635 kPa, the value of ε_g increase by a factor of 2.24. This observation indicates the compressible nature of the gel layer over the range of operating pressure considered here. At higher pressure, the compressibility of gel layer becomes less compared to lower pressure. Since α typically increases with pressure (resulting in gel compaction), gel porosity shows a decreasing trend with increase in pressure. The evaluated optimum values of α are correlated with operating pressure by a simple relation,

$$\alpha = \alpha_0 \left(\Delta P\right)^{0.784} \qquad (2.17)$$

where, $\alpha_0 = 1.313 \times 10^{15}$ m/kg and ΔP is in Pa.

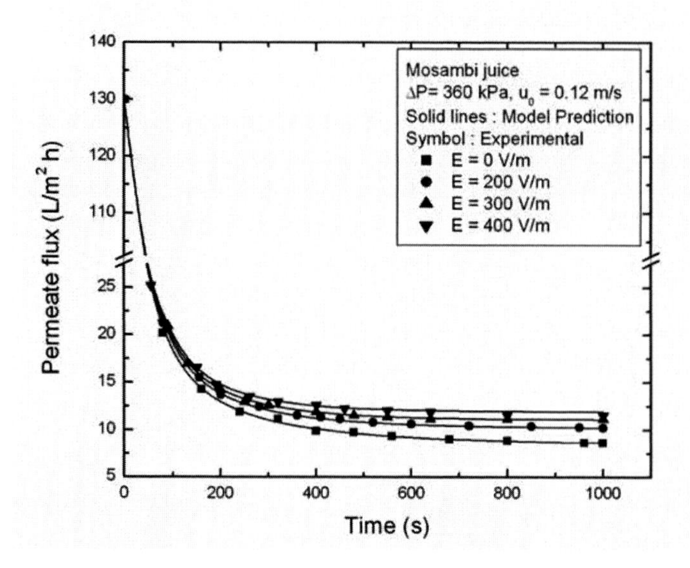

Figure 2.11. Variation of permeate flux with time for different electric field during clarification of mosambi juice.

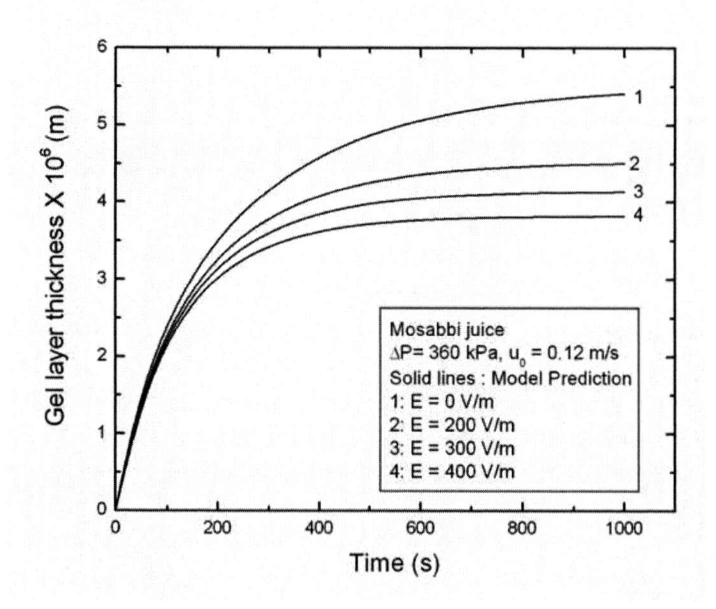

Figure 2.12. Variation of gel layer thickness with time for different electric field during clarification of mosambi juice.

Table 2.7. Variation of specific gel layer resistance (α) with operating pressure for mosambi juice

Transmembrane pressure, ΔP (kPa)	Specific gel layer resistance, α (m/kg)
220	15.6×10^{15}
360	23.4×10^{15}
500	31.1×10^{15}
635	35.0×10^{15}

Effect of Pulsed Electric Field

The variation of permeate flux and gel layer thickness with time for different pulse (on-time/off-time ratio) ratio at a transmembrane pressure of 360 kPa and cross flow velocity of 0.12 m/s during clarification of mosambi juice are shown in Figures 2.13 and 2.14 respectively. The symbols are for the experimental permeate flux, whereas, the solid lines are model predicted permeate flux and the dashed lines are estimated gel layer thickness. For a given set of operating conditions, the value of specific gel resistances (α) is obtained by minimizing the root mean square error between the measured and calculated flux values. For example, at a transmembrane pressure of 360 kPa, cross flow velocity of 0.12 m/s, and electric field of 400 V/m, the optimum value of α is found to be 2.34×10^{16} m/kg. It is also observed that the value of α is independent of pulse ratio. It is evident from the figure that permeate flux declines rapidly as the gel layer grows and reaches a steady value. The same trend can be observed for all pulse ratios. In this study, the off-time is always fixed at 1s. It can clearly be seen that an increase in pulse ratio (i.e., increase in on-time) results in a decrease in the estimated gel layer thickness (less resistance to flow) and associated with increase in flux. For a fixed electric field, with increase in pulse ratio i.e., on-time, charged pectin molecules move at a faster rate from the membrane surface. This restricts the formation of gel-type layer over the membrane surface and leads to an enhancement of permeate flux. For example, at 400 V/m, a transmembrane pressure of 360 kPa, a cross flow velocity of 0.12 m/s, and at a pulse ratio of 1:1 (i.e., 1 s on followed by 1 s off time), permeate flux declines from 130 L/m^2 h to 10.4 L/m^2 h and gel layer thickness grows up to 4.45 μm at the end of operation whereas, at a pulse ration of 3:1(i.e., 3 s on followed by 1 s off time), the permeate flux declines to 11.35 L/m^2 h with a gel layer thickness of 4.04 μm keeping other operating conditions unaltered.

2.4.2. Analysis of Steady State Flux

2.4.2.1. Theoretical Aspect

The steady state solute mass balance in the rectangular channel within the concentration boundary layer is given by

$$u\frac{\partial c}{\partial x} + (v + v_e)\frac{\partial c}{\partial y} = D\frac{\partial^2 c}{\partial y^2} \tag{2.18}$$

where, v_e is the electrophoretic velocity which can be obtained from Eq. (2.8). The electrophoretic mobility (u_e) can be obtained from Eq. (2.9). The diffusion coefficient (D) is given by Einstein equation (Eq. (2.11)). Assuming hydrodynamic velocity profile to be fully developed, the x-component velocity becomes [29]

$$u = \frac{3}{2}u_o\left[1 - \left(\frac{y\text{-}h}{h}\right)^2\right] \tag{2.19}$$

The above equation of velocity is obtained assuming the permeation velocity of solvent through the membrane is small enough so that the x-component velocity profile is not distorted [30]. Within the thin concentration boundary layer, the term $\frac{y^2}{h^2}$ can be neglected and the x-component velocity profile can be simplified as,

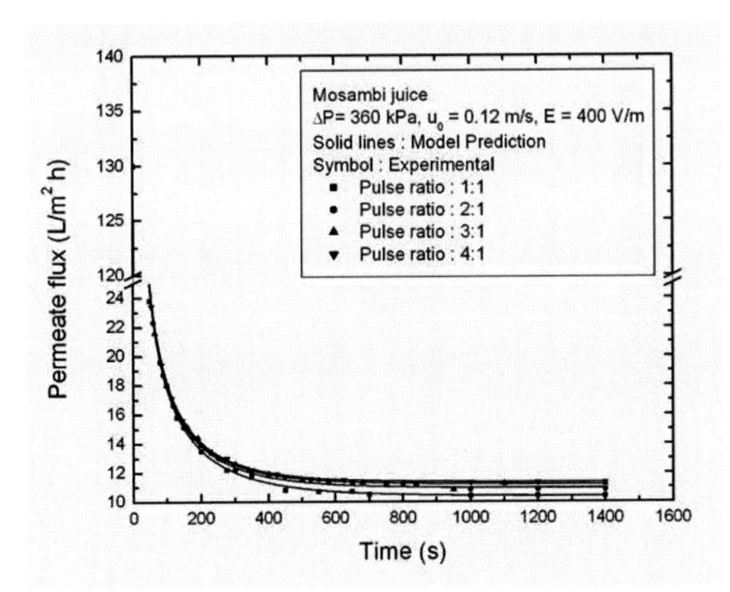

Figure 2.13. Variation of permeate flux with time for different pulse ratio during clarification of mosambi juice.

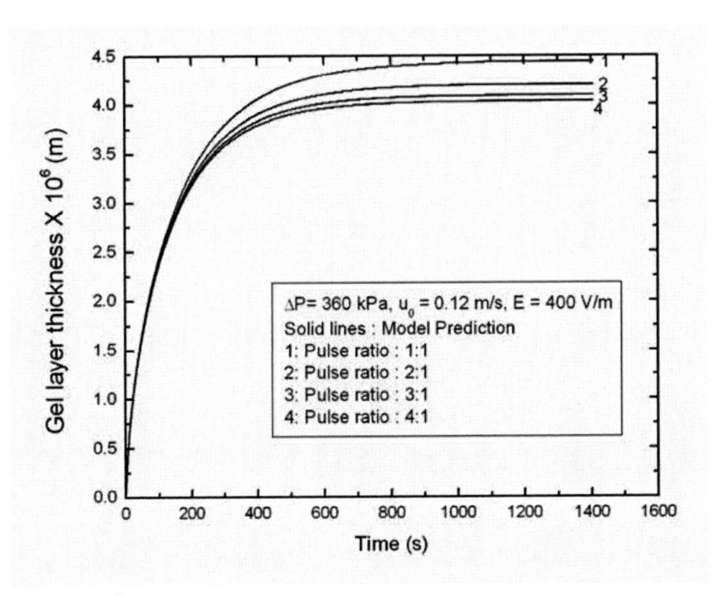

Figure 2.14. Variation of gel layer thickness with time for different pulse ratio during clarification of mosambi juice.

$$u = \frac{3u_0 y}{h} \tag{2.20}$$

Since the thickness of concentration boundary layer is extremely small as $\delta_c \propto \frac{1}{Sc^{\frac{1}{3}}}$ and Schmidt number is quite large due to lower diffusivity of pectin molecule, the y-component velocity within concentration boundary layer is approximated as,

$$v = - v_w \tag{2.21}$$

The boundary conditions of Eq. (2.18) are,

$$\text{at } x=0,\ c = c_0 \text{ at } y=\delta_c,\ c = c_0 \tag{2.23}$$

where, δ_c is the thickness of the concentration boundary layer.

The boundary condition at the gel-solution interface can be written as (equality of convective movement of the solutes towards the membrane to the diffusive and electrophoretic movement away from the membrane),

$$\text{at } y=0,\ D\frac{\partial c}{\partial y} + (v_w - v_e)c_g = 0 \tag{2.24}$$

Inserting the velocity profiles i.e., Eqs. (2.20) and (2.21), the governing mass balance equation (Eq.(2.18)) can be non-dimensionalized as,

$$Ay^* \frac{\partial c^*}{\partial x^*} + \left(\frac{Pe_e\text{-}Pe_w}{4}\right)\frac{\partial c^*}{\partial y^*} = \frac{\partial^2 c^*}{\partial y^{*2}} \tag{2.25}$$

where $x^* = \frac{x}{L};\ c^* = \frac{c}{c_0};\ A = \frac{3}{16}\left(ReSc\frac{d_e}{L}\right)$

$$Pe_e = \frac{v_e d_e}{D}, \quad Pe_w = \frac{v_w d_e}{D}, \quad y^* = \frac{y}{h}$$

The initial and boundary conditions are non-dimemsionalized as,

$$\text{at } x^* = 0, \quad c^* = 1 \tag{2.26}$$

$$\text{at } y^* = \delta_c^*, \quad c^* = 1 \tag{2.27}$$

$$\text{and at } y^* = 0, \quad \frac{\partial c^*}{\partial y^*} + \left(\frac{Pe_w - Pe_e}{4}\right) c_g^* = 0 \tag{2.28}$$

Using the integral approach, the following concentration profile is assumed:

$$c^* = \frac{c}{c_0} = a_1 + a_2\left(\frac{y^*}{\delta_c^*}\right) + a_3\left(\frac{y^*}{\delta_c^*}\right)^2 \tag{2.29}$$

where, $\delta_c^* = \dfrac{\delta_c}{h}$

Eq. (2.29) satisfies the following boundary conditions:

$$\text{at } y^* = \delta_c^*, \quad c^* = 1 \tag{2.30}$$

$$\text{at } y^* = \delta_c^*, \quad \frac{\partial c^*}{\partial y^*} = 0 \tag{2.31}$$

$$\text{at } y^* = 0, \quad c^* = \frac{c_g}{c_0} = c_g^* \tag{2.32}$$

where, c_g is the gel concentration.

With the help of the above boundary conditions Eq. (2.29) can be written as,

$$c^* = c_g^* - 2(c_g^* - 1)(\frac{y^*}{\delta_c^*}) + (c_g^* - 1)(\frac{y^*}{\delta_c^*})^2 \tag{2.33}$$

Substituting the partial derivative of c^* with respect to x^* and y^* from Eq. (2.33) into the non-dimentionalized governing equation (Eq.(2.25)), the following equation is obtained;

$$A\left(\frac{y^*}{\delta_c^*}\right)^2\left(1-\frac{y^*}{\delta_c^*}\right)\frac{d\delta_c^*}{dx^*}+\left(\frac{Pe_w - Pe_e}{4\delta_c^*}\right)\left(1-\frac{y^*}{\delta_c^*}\right)=\frac{1}{\delta_c^{*2}} \tag{2.34}$$

Multiplying both sides of Eq. (2.34) by dy^* (taking zeroth moment) and then integrating across the boundary layer thickness, i.e. from 0 to δ_c^* the following expression is obtained:

$$A\frac{d\delta_c^*}{dx^*}\int_0^{\delta_c^*}\left(\frac{y^{*2}}{\delta_c^{*2}}-\frac{y^{*3}}{\delta_c^{*3}}\right)dy^* + \left(\frac{Pe_w - Pe_e}{4}\right)\int_0^{\delta_c^*}\left(\frac{1}{\delta_c^*}-\frac{y^*}{\delta_c^{*2}}\right)dy^* = \frac{1}{\delta_c^{*2}}\int_0^{\delta_c^*}dy^* \tag{2.35}$$

After integration of Eq. (2.35), the following expression is obtained:

$$\frac{A\,\delta_c^{*2}}{12}\frac{d\delta_c^*}{dx^*}+\frac{(Pe_w - Pe_e)\,\delta_c^*}{8}=1 \tag{2.36}$$

Following equation is obtained by substituting the partial derivative of c^* with respect to y^* from Eq. (2.33) into the Eq. (2.28).

$$(Pe_w - Pe_e)\,\delta_c^* = \frac{8(c_g^*-1)}{c_g^*} \tag{2.37}$$

Substituting Eq. (2.37) into Eq. (2.36) and after simplification the following expression is obtained:

$$\frac{A\delta_c^{*2}}{12}\frac{d\delta_c^*}{dx^*} = \frac{1}{c_g^*} \tag{2.38}$$

at $x^* = 0, \ \delta_c^* = 0$ (2.39)

The solution of the above expression can finally be written as,

$$\delta_c^* = \left(\frac{36x^*}{Ac_g^*}\right)^{\frac{1}{3}} \tag{2.40}$$

Substituting the Eq. (2.40) into Eq. (2.37) the following expression is finally obtained:

$$Pe_w = Pe_e + \frac{8(c_g^*-1)}{c_g^{*\frac{2}{3}}}\left(\frac{A}{36x^*}\right)^{\frac{1}{3}} \tag{2.41}$$

The above solution provides the variation of permeate flux along the length of the channel for different electric field, feed concentration and cross flow velocity.

Finally, length averaged permeate flux is obtained by integrating the Eq. (2.41) as,

$$\overline{Pe_w} = \int_0^1 Pe_w(x^*)\,dx^* \tag{2.42}$$

$$\overline{Pe_w} = Pe_e + 2.1\left(ReSc\frac{d_e}{L}\right)^{\frac{1}{3}}\frac{(c_g^*-1)}{c_g^{*\frac{2}{3}}} \tag{2.43}$$

Eq. (2.43) can be simplified as,

$$\overline{\mathrm{Pe}_w} = \mathrm{Pe}_e + 2.1\left(\mathrm{ReSc}\frac{d_e}{L}\right)^{\frac{1}{3}} \ln c_g^* \ \text{ when, } c_g^* < e^3 \tag{2.44}$$

2.4.2.2 Analysis of Steady State Flux of Synthetic Juice

Effect of Constant Electric Field

Figure 2.15 and 2.16 represent the variations of permeate flux with applied external d.c. electric field for different values of feed concentration and transmembrane pressure. The symbols are experimental data and the solid lines are model predictions. It may be observed from the figure that permeate flux increases almost linearly with applied electric field. When external d.c. electric field is applied with appropriate polarity, the electrophoretic movement of pectin towards the positive electrode causes reduction of deposition on the membrane surface and gel layer formation is restricted. As a result, permeate flux increases. From the Figure 2.6 it is observed that the permeate flux increases by a factor 2.78 with an increase in electric field from 0 to 800 V/m for a fixed feed concentration (1 kg/m^3 +14 Brix), pressure (220 kPa) and cross-flow velocity (0.12 m/s). Similar trends are observed in case of other cross flow velocity i.e., at 0.09 m/s, the permeate flux increases by a factor 2.87 with the same increment of electric field (Figure 2.16). Moreover, from the results shown in this figure, lower feed concentration results in higher permeate flux due to less concentration polarization on the membrane surface with other operating conditions held constant.

Effect of Transmembrane Pressure

During gel-layer governed ultrafiltration, for a fixed values of feed concentration, cross flow velocity and electric field, the permeate flux is, by definition, independent of pressure. The enhanced driving force for solvent flux due to increase in pressure is fully compensated by the resistance offered by the growing gel layer on the membrane surface and/or due to compaction of the gel layer in an ideal situation. However, weak pressure dependence is observed in our experimental results, especially at higher values of operating pressure (360 kPa). Nevertheless, the average deviation is of the order of 10% only. This may be due to departure from the ideal gel layer controlled operation (as described above).

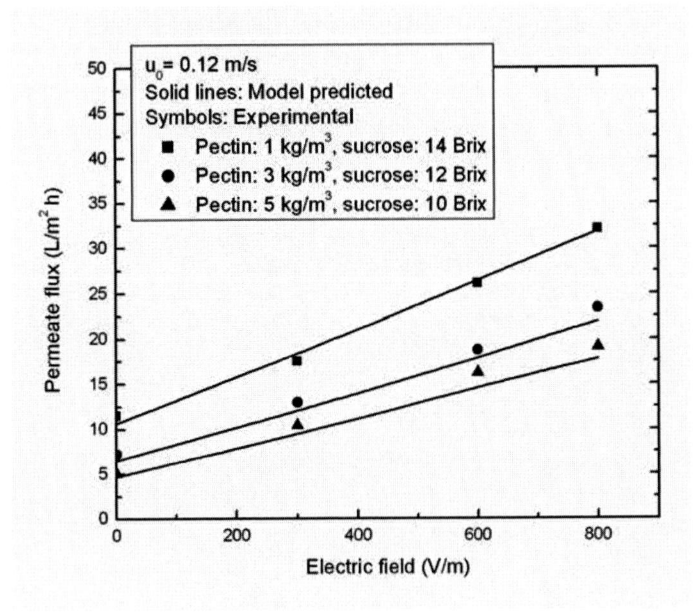

Figure 2.15. Variations of permeate flux with electric field for different feed concentration at a pressure of 220 kPa and cross velocity of 0.12 m/s.

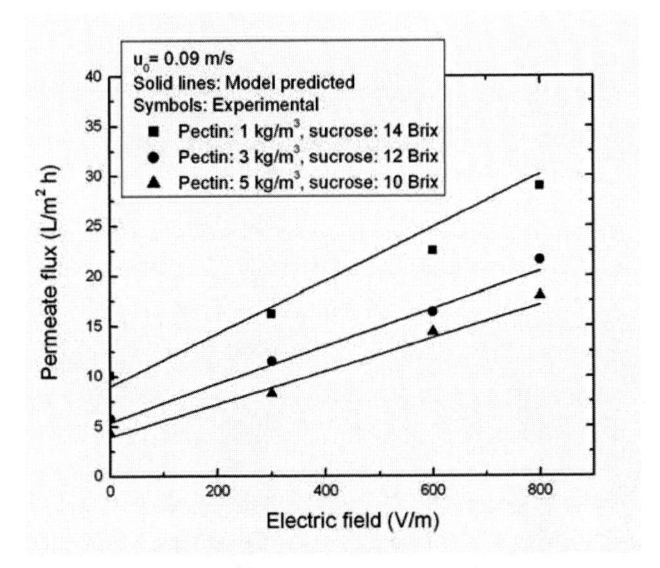

Figure 2.16. Variations of permeate flux with electric field for different feed concentration at a pressure of 220 kPa and cross velocity of 0.09 m/s.

This trend is more clear if the permeate fluxes are plotted as a function of electric field at different operating pressures (Figures 2.17 and 2.18). The departure from pressure independence of permeate flux are more pronounced for higher transmembrane pressure (as is evident from the slight positive slopes of the four solid lines connecting the experimental data points).

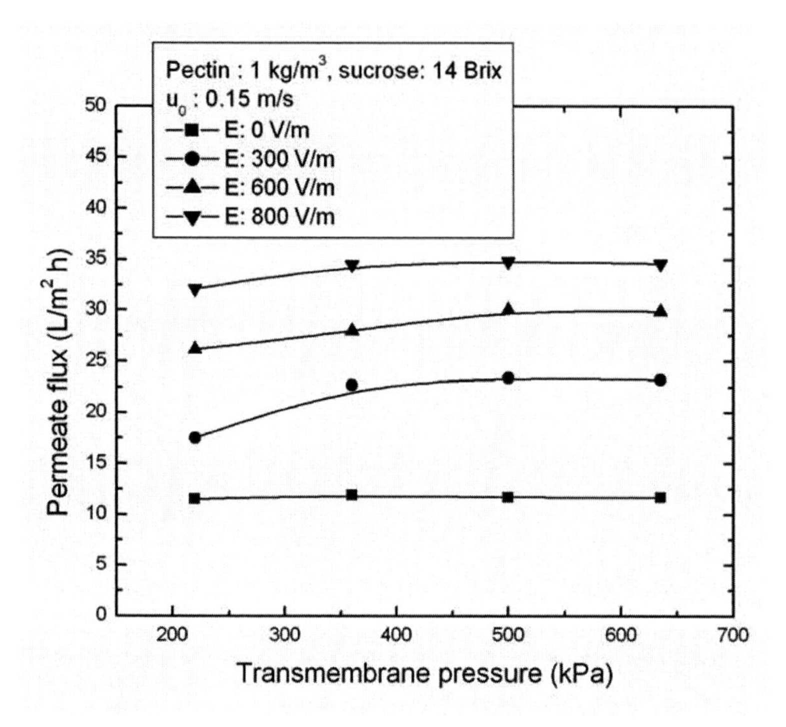

Figure 2.17. Variations of experimental permeate flux with transmembrane pressure difference for different electric field. The solid lines are only guides for the reader.

Effect of Cross Flow Velocity

The variations of permeate flux with cross flow velocity for different electric fields are presented in Figure 2.20. From the results shown in this figure, it is clear that for a fixed electric field permeate flux increases with increase in cross flow velocity. At higher cross flow velocity, gel layer thickness on the membrane surface is lower due to forced convection. Therefore, the back diffusion from membrane surface to the bulk increases as the concentration gradient from the membrane surface to the bulk increases. This leads to an increase of permeate flux at higher cross flow velocity. For example, for feed concentration (1 kg/m^3 +14 Brix), at 600 V/m, permeate flux increases from 22.5 L /m^2 h to 27.0 L /m^2 h with an increase of cross flow velocity from 0.09 m/s to 0.18 m/s. The calculated values of permeate flux using the developed model are also presented by the solid lines in Figure 2.19. It is clear from this figure that the model predictions are within 10% of the experimental data.

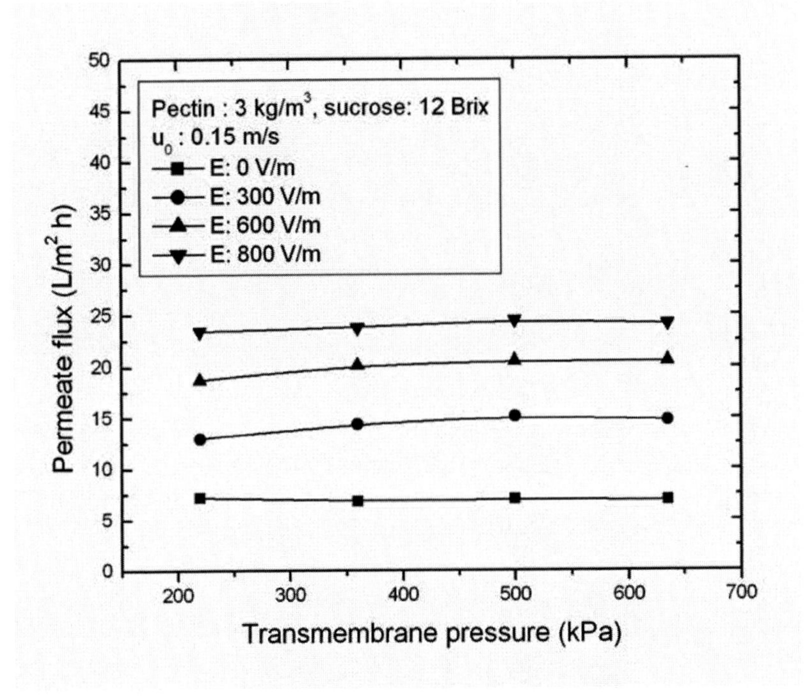

Figure 2.18. Variations of experimental permeate flux with transmembrane pressure difference for different electric field. The solid lines are only guides for the reader.

Variation of Permeate Flux with Axial Position

Figures 2.20 to 2.24 show the variation of permeate flux along the dimensionless length of the ultrafiltration channel for various feed concentration and cross flow velocity. It is observed from the figures that during the initial part of the channel, flux decline is very rapid and then gradual for the later part of the channel. This may be due to the sluggish development of concentration boundary layer in the downstream of the channel because of the forced convection imposed by the cross flow. For example, for feed concentration (1 kg/m^3+14 Brix), at 800 V/m and a cross flow velocity of 0.15 m/s, permeate flux decreases from 77.4 L /m^2 h to 33.6 L /m^2 h along the length of the channel. At any position of the channel an increase in flux is increased with increase in cross flow velocity and

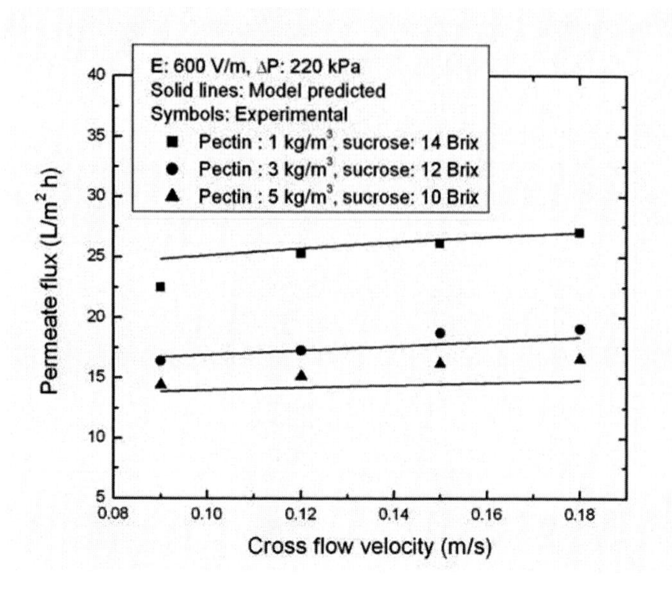

Figure 2.19. Variations of permeate flux with cross flow velocity for different electric field at a fixed pressure of 220 kPa and an electric field of 600 V/m.

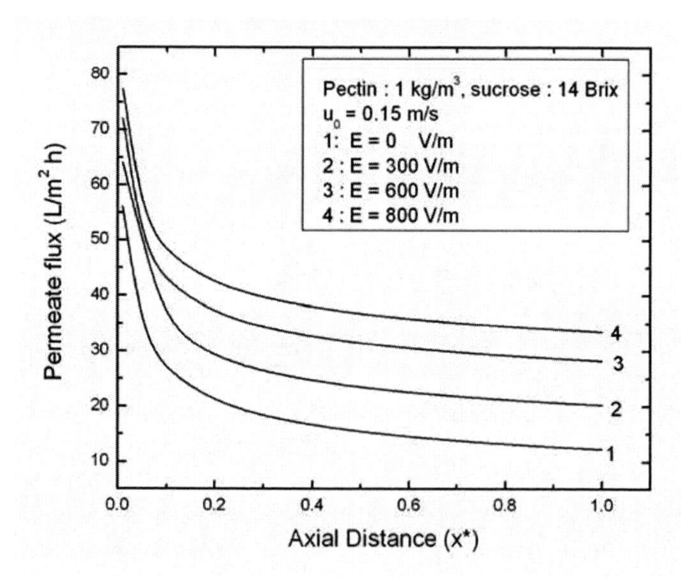

Figure 2.20. Variations of permeate flux along the channel length for different values of feed concentration of pectin: 1 kg/m^3, sucrose: 14 Brix and cross flow velocity of 0.15 m/s.

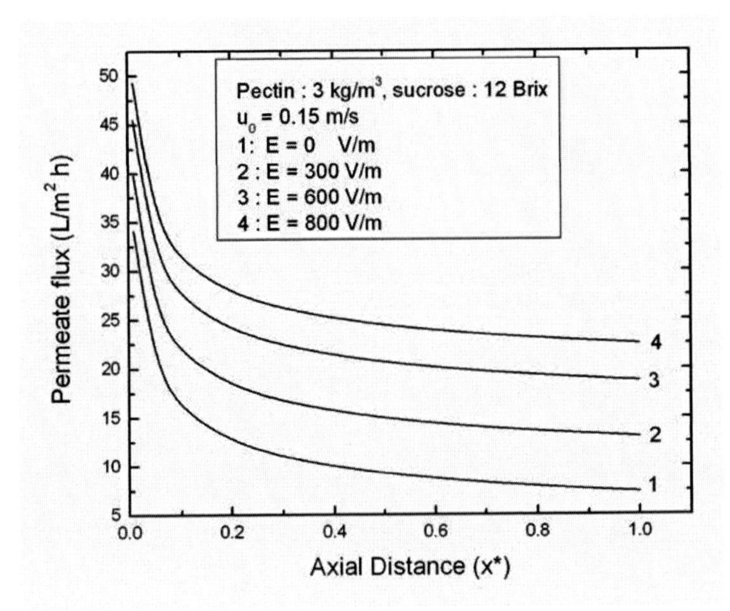

Figure 2.21. Variations of permeate flux along the channel length for different values of feed concentration of pectin: 3 kg/m^3, sucrose: 12 Brix and cross flow velocity of 0.15 m/s.

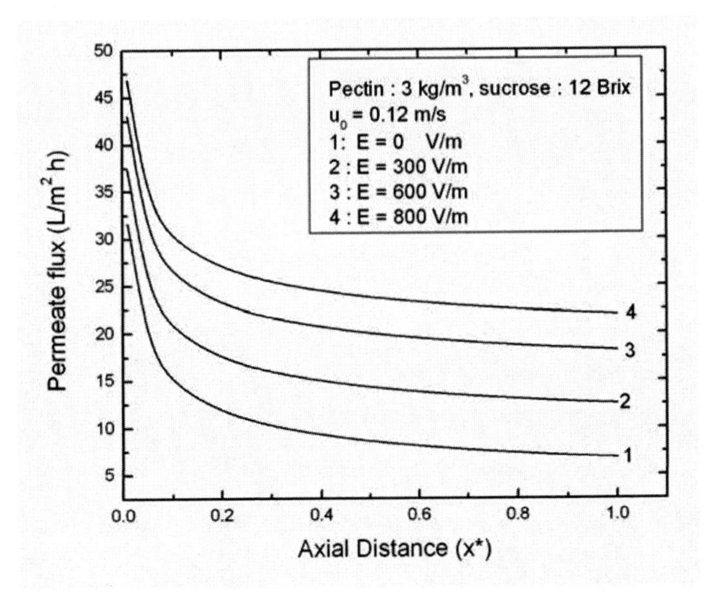

Figure 2.22. Variations of permeate flux along the channel length for different values of feed concentration of pectin: 3 kg/m^3, sucrose: 12 Brix and cross flow velocity of 0.12 m/s.

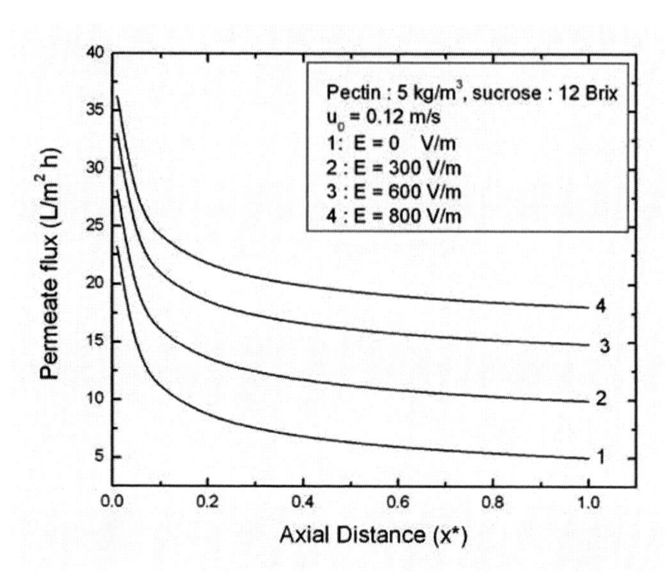

Figure 2.23. Variations of permeate flux along the channel length for different values of feed concentration of pectin: 5 kg/m^3, sucrose: 10 Brix and cross flow velocity of 0.12 m/s.

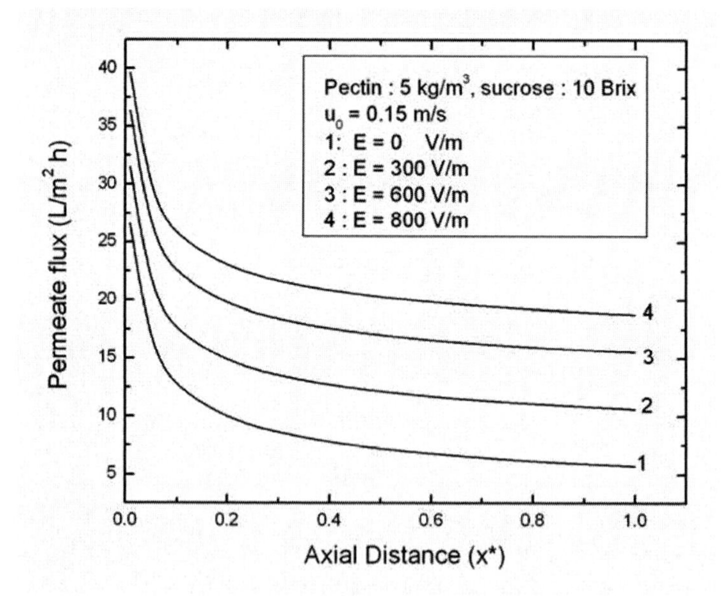

Figure 2.24. Variations of permeate flux along the channel length for different values of feed concentration of pectin: 5 kg/m^3, sucrose: 10 Brix and cross flow velocity of 0.15 m/s.

decrease in feed concentration. For example, at a cross flow velocity of 0.12 m/s permeate flux is about 22.0 L /m^2 h for feed concentration (3 kg/m^3+12 Brix) and that is for feed concentration (5 kg/m^3+10 Brix) is about 18.07 L /m^2 h at the end of the channel; whereas, at a cross flow velocity of 0.15 m/s, permeate flux is about 22.55 L /m^2 h for feed concentration (3 kg/m^3+12 Brix).

Effect of Pulsed Electric Field

Effect of Pulse Ratio

Figures 2.25 and 2.26 show the variation in steady state average permeate flux and gel layer thickness with pulse (on time/off time) ratio for various electric field at a transmembrane pressure of 360 kPa and a cross flow velocity of 0.12 m/s during clarification of synthetic juice respectively. In Figure 2.15, the symbols are for the experimental permeate flux, whereas, the solid lines are model predicted permeate flux. As can be seen from the figure, the predictions are very close to the experimental values of permeate flux. In fact it can be seen that the model predictions are within ±5% of the experimental results. The off-time is always fixed at 1s. It can clearly be seen that an increase in electric field results in a decrease in the estimated gel layer thickness (less resistance to flow) and associated increase in flux. With increase in electric field, movement of charged pectin molecules away from the membrane surface increases. This restricts the formation of gel-type layer over the membrane surface and leads to an enhancement of permeate flux. For example, for a feed consisting of 3 kg/m^3 Pectin and 12 Brix of Sucrose, at a transmembrane pressure of 360 kPa, cross flow velocity of 0.12 m/s, pulse ratio of 3:1 and electric field of 2000 V/m, the permeate flux is augmented by a factor of 6.0 compared to zero electric field and corresponding values of gel layer thickness decreases from 10.68 μm to 1.3 μm (Figure 2.26). Experimental results confirm that for a given set of operating conditions, 94.5-97.8% of the maximum permeate flux (electric field always on) is obtained at a pulse ratio of 3:1(i.e., 3 s on followed by 1 s off time). Further increase in pulse ratio does not improve the permeate flux significantly and similar trends are observed for all electric fields.

Figures 2.27 and 2.28 illustrate the variation in steady state average permeate flux and gel layer thickness with pulse ratio for various feed concentrations (mixture of pectin and sucrose) at a transmembrane pressure of 360 kPa and a cross flow velocity of 0.12 m/s during clarification of synthetic juice respectively. In Figure 2.17, the symbols are for the experimental permeate flux, whereas, the solid lines are model predicted permeate flux. It may be observed that the predicted flux values are within ±5% to the experimental results. From the figure it is clear that almost same level of the maximum achievable permeate flux (electric field always on) can be obtained at the pulse ratio of 3:1. As can be seen from the figure, permeate flux decreases with increase in feed concentration while other operating conditions are kept unchanged. At higher feed concentration, concentration polarization becomes more severe and results in higher value of gel layer thickness. This increases the resistance to solvent flow through the membrane and hence permeate flux decreases at higher feed concentration. For example, at a transmembrane pressure of 360 kPa, cross flow velocity of 0.12 m/s, pulse ratio of 3:1 and electric field of 800 V/m, with increase in feed concentration from a mixture consisting of 1 kg/m^3 Pectin and of 14 Brix Sucrose to a mixture consisting of 5 kg/m^3 Pectin and of 10 Brix Sucrose, the permeate flux declines from 31.1 L/m^2 h to 15.9 L/m^2 h and corresponding values of gel layer thickness over the membrane surface increases from 1.56 μm to 5.1 μm (Figure 2.28). It may also be observed that application of electric field is more effective at higher feed concentrations. This result is found to be consistent with that reported in the literature [31].

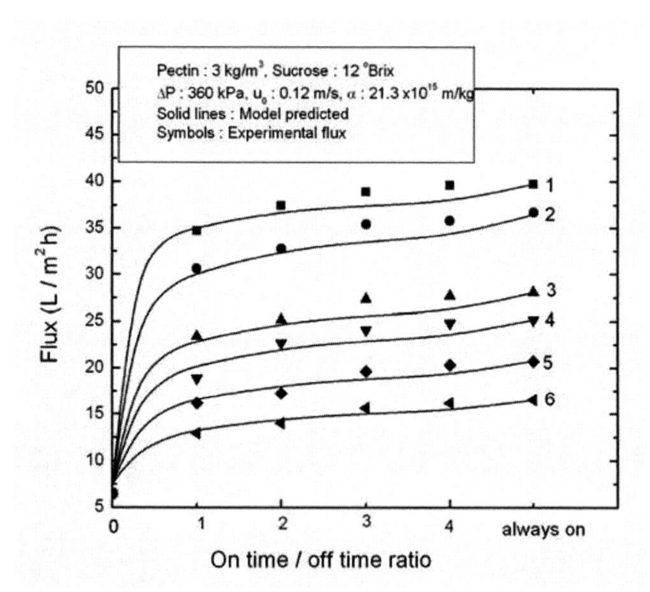

Figure 2.25. Variation in steady state average permeate flux with pulse ratio for various electric fields during clarification of synthetic juice. ■, Curve 1 for E = 2000 V/m; ●, Curve 2 for E = 1600 V/m; ▲, Curve 3 for E = 1200 V/m; ▼, Curve 4 for E = 1000 V/m; ♦, Curve 5 for E = 800 V/m; ◄, Curve 6 for E = 600 V/m.

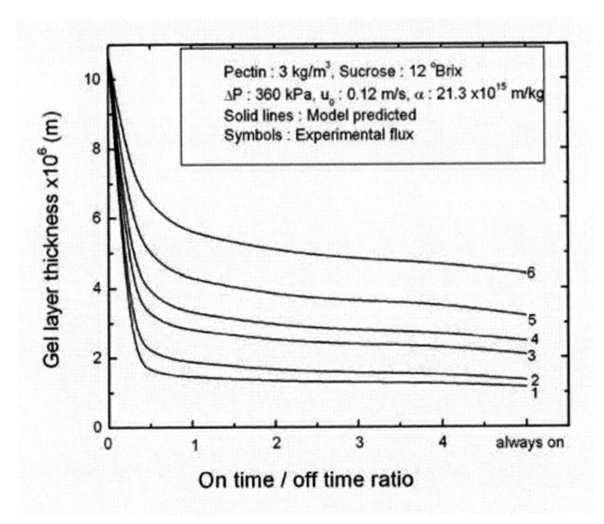

Figure 2.26. Variation in steady gel layer thickness with pulse ratio for various electric fields during clarification of synthetic juice. Curve 1 for E = 2000 V/m; Curve 2 for E = 1600 V/m; Curve 3 for E = 1200 V/m; Curve 4 for E = 1000 V/m; Curve 5 for E = 800 V/m; Curve 6 for E = 600 V/m.

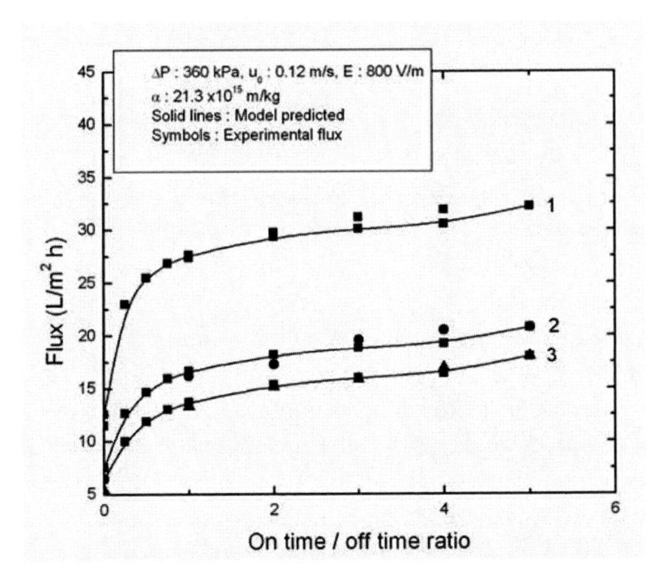

Figure 2.27. Variation in steady state average permeate flux with pulse ratio for various feed concentration during clarification of synthetic juice. ■, Curve 1 for c_o = 1 kg/m^3, 14 Brix; ♦, Curve 2 for c_o = 3 kg/m^3, 12 Brix; ▲, Curve 3 for c_o = 5 kg/m^3, 10 Brix.

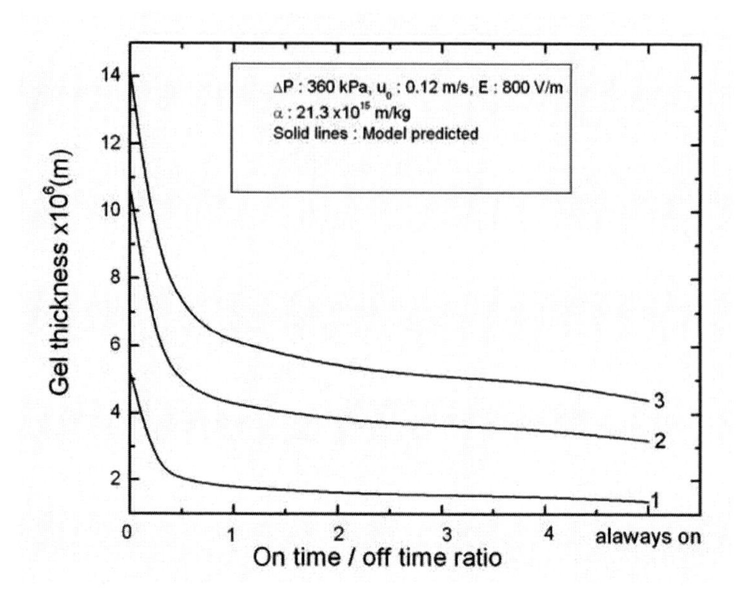

Figure 2.28. Variation in steady state gel layer thickness with pulse ratio for various feed concentration during clarification of synthetic juice. Curve 1 for c_o = 1 kg/m^3, 14 Brix; Curve 2 for c_o = 3 kg/m^3, 12 Brix; Curve 3 for c_o = 5 kg/m^3, 10 Brix.

Effect of Cross Flow Velocity

Variations in steady state average permeate flux with cross flow velocity at a transmembrane pressure of 360 kPa and pulse ratio of 3:1 during clarification of synthetic juice are reported in Figure 2.29. The symbols are the experimental data and solid lines are model predictions and they are within ±5% of each other. It may be observed from the figure that with increase in cross flow velocity, permeate flux increases as the gel thickness over the membrane surface is reduced by enhanced forced convection. For example, for a specific feed concentration (Pectin: 3 kg/m^3, Sucrose: 12 Brix), transmembrane pressure of 360 kPa, pulse ratio of 3:1 and at an electric field of 800 V/m, an increase in cross flow velocity from 0.09 m/s to 0.18 m/s leads to an enhancement of permeate flux by about 17.7 % whereas, at 2000 V/m, this enhancement is about 7.8 %. This also indicates that the effect of cross flow velocity is more pronounced at 800 V/m compared to 2000 V/m. At a higher electric field the thickness of gel-type layer is already thin and an increase in cross flow velocity may not have an appreciable effect.

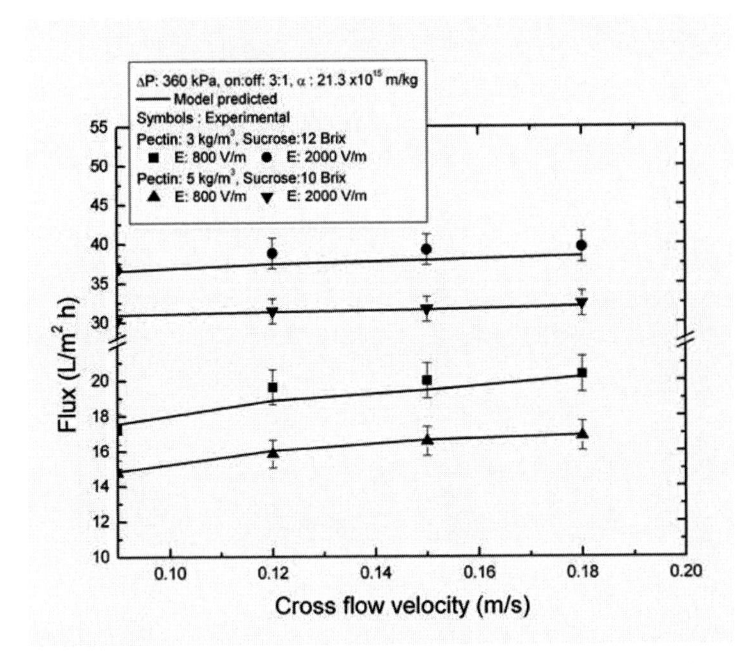

Figure 2.29. Variation in steady state average permeate flux with cross flow velocity for different electric field and a fixed pulse ratio of 3:1 during clarification of synthetic juice.

Estimation of Electric Power Consumption per Unit Volume of Permeate

Total energy consumption is the summation of energy required to run the pump and energy to supply the external d.c. electric field. Pump-dissipated power is directly proportional to the pressure drop and can be expressed as the product of pressure drop and feed flow rate. Electric power is related to the applied voltage. Hence, the total specific energy consumption i.e., energy consumption per unit volume of permeate can be written as,

$$E_{total} = E_{pump} + E_{electical} \tag{2.45}$$

$$E_{total} = \frac{Q\,\Delta P_{pump}}{v_w A' \,\eta_{pump}} + \frac{V\,I\,N}{v_w A' \,\eta_{d.c\ power\ supply}} \tag{2.46}$$

where, Q, ΔP_{module}, V, I, N, v_W, A, η_{pump}, $\eta_{d.c.\ power\ supply}$ are the feed flow rate, pressure drop, applied voltage, current, fraction of on time, permeate flux, membrane area, efficiency of pump and d.c. power supply respectively.

The variation of pH and electric power consumption of the permeate at various electric fields and pulse ratios during clarification of synthetic juice (Pectin: 3 kg/m^3, Sucrose: 12 Brix) and mosambi juice at a transmembrane pressure of 360 kPa and cross flow velocity of 0.12 m/s are shown in Figures 2.30 and 2.31 respectively. It is generally observed that power consumption and pH of the permeate increases with increase in electric field as well as with pulse ratio.

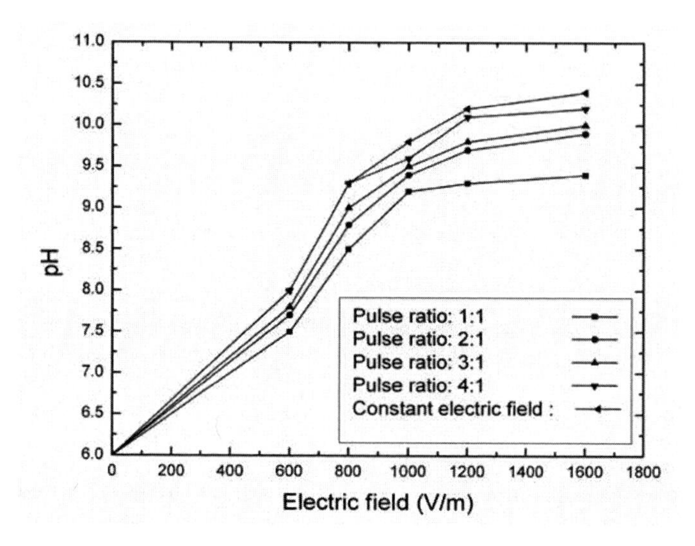

Figure 2.30. Variation of pH of the permeate with d.c. electric field at various pulse ratios at a pressure of 360 kPa and cross flow velocity of 0.12 m/s during clarification of synthetic juice (pectin: 3 kg/m^3, sucrose: 12 Brix).

As observed in Figure 2.27, for a given electric field, nearly the maximum flux value is achieved at a pulse ratio of 3:1 which corresponds to (27-33) % less energy consumption compared to the constant electric field. From experimental results, it can be seen that pH of the permeate is maximum at constant electric field compared to pulsed electric field. At constant field, electrode reactions resulting in evolution of oxygen (O_2) and hydronium ions (H_3O^+) at the anode and hydrogen (H_2) and hydroxyl ions (OH^-) at the cathode are more pronounced compared to that with the pulsed electric field, these hydroxyl ions (OH^-) are carried with the permeate stream and may result an increase in pH of permeate.

Electrolytic gases are released near the membrane resulting in a reduction of permeate flux. For example, pH of the permeate is increased from 6.0 to 9.3 with increase in constant electric field from 0 V/m to 800 V/m, whereas, with a pulse ratio of 3:1, the pH of the permeate increases up 9.0.

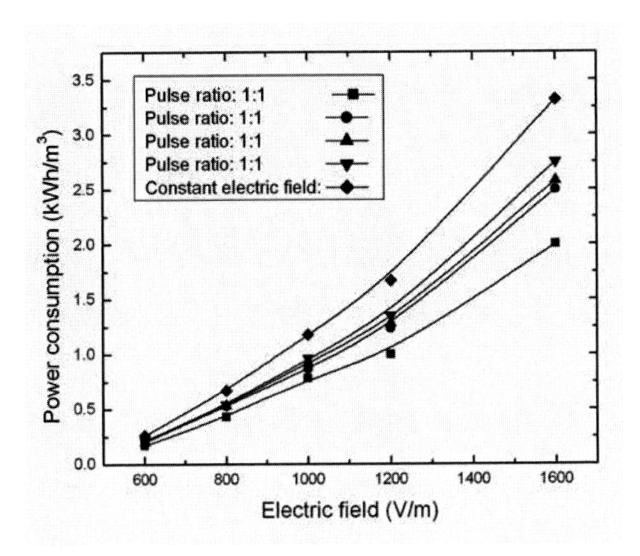

Figure 2.31. Variation of electric power consumption per unit volume of the permeate with d.c. electric field at various pulse ratios at a pressure of 360 kPa and cross flow velocity of 0.12 m/s during clarification of synthetic juice (pectin: 3 kg/m^3, sucrose: 12 Brix).

Table 2.8. Variation of energy ratio of pulse to constant electric field with pulse ratio for different electric fields

ΔP: 360 kPa, u_0 : 0.12 m/s	E = 200 V/m	E = 300 V/m	E = 400 V/m
Pulse	$E_{pulse}/E_{constant}$	$E_{pulse}/E_{constant}$	$E_{pulse}/E_{constant}$
0	0	0	0
1	0.25	0.34	0.41
2	0.38	0.49	0.56
3	0.56	0.58	0.67
4	0.67	0.71	0.76
Constant electric field	1.0	1.0	1.0

Table 2.8 illustrates the variation of energy ratio of pulse to constant electric field with pulse ratio for different electric field. It is evident from the figure that with increase in pulse ratio, the dimensionless energy, $E_{pulse}/E_{constant}$, increases for a fixed value of transmembrane pressure and cross flow velocity. It can be observed from experimental results that at an electric field of 400 V/m, at a pressure of 360 kPa, a cross flow velocity of 0.12 m/s and at the pulse ratio of 3:1, the dimensionless energy, $E_{pulse}/E_{constant}$ is about 0.67. This indicates that during pulsed electro-ultrafiltration, about 97.5% of the maximum attainable permeate flux (electric field always on) is achieved (Figure 2.27) with saving of 33% electrical energy.

Table 2.9 represents the variations of energy consumption per unit volume of permeate with external applied constant d.c. electric field for a fixed pressure of 360 kPa and a cross flow velocity of 0.12 m/s. It is observed from the figure that with increase in electric field, energy consumption due to applied d.c. electric field increases but the energy consumed by the pump decreases. However, total energy consumption per unit volume of permeate decreases. This can be explained by the fact that for a fixed pressure and cross flow velocity, with increase in electric field, energy consumption due to d.c electric field is increases. Though at the same time, with increase in electric field, permeate flux increases due to decrease of gel layer thickness (migration of gel forming materials away from the membrane surface), the overall increase of energy consumption per unit of permeate volume due to d.c electric field ($E_{electrical}$) is observed. For example, at 400 V/m, the value of $E_{electrical}$ is about 1.72 kWh/m^3 of permeate. For the calculation of pump energy consumption, there is no contribution of d.c. electric field. Hence, consumption of pump energy per unit volume of permeate (E_{pump}) decreases with increase of electric field as permeate flux increases with increase in electric field. For example, E_{pump} decreases from 123.8 to 92.85 kWh/m^3 of permeate with increase in electric field from 0 to 400 V/m. Moreover, since the contribution of $E_{electrical}$ to the calculation of E_{total} is less than 2%, with increase in electric field, total energy consumption per unit volume of permeate (E_{total}) decreases. Therefore, the energy consumption due to application of external electric field insignificant compared to that for running the pump.

Table 2.9. Variation of energy consumption with electric field.

E (V/m)	E_{pump} (kWh/m^3)	$E_{electrica}$ (kWh/m^3)	E_{total}(kWh/m^3)
0	123.8	0	123.8
200	104.4	0.15	104.55
300	96.5	0.63	96.65
400	92.85	1.73	94.58

Table 2.10 illustrates the variations of energy consumed by pump per unit volume of permeate for different transmembrane pressures and cross flow velocities at an electric field of 400 V/m and pulse ratio of 3:1. It is evident from the figure that energy consumed by the pump per unit volume of permeate increases with increase in both transmembrane pressure as well as cross flow velocity. This is because, increase in cross flow velocity and transmembrane pressure lead to enhancement of power consumption of pump as it is directly proportional to both cross flow velocity and pressure drop. For example, at a cross flow velocity of 0.12 m/s, with increase in transmembrane pressure from 220 kPa to 635 kPa, pump energy consumption increases from 63.7 kWh/m^3 to 158.8 kWh /m^3 whereas, at a transmembrane pressure of 360 kPa, with increase in cross flow velocity from 0.09 m/s to 0.18 m/s, pump energy consumption increases from 76.83 kWh /m^3 to 127.32 kWh /m^3.

Table 2.10 also shows the variations of energy consumed by d.c. electric field per unit volume of permeate for different transmembrane pressures and cross flow velocities at an electric field of 400 V/m and a pulse ratio of 3:1 are presented in Figure 2.24.

Table 2.10. Variation of pump energy and electrical energy for different transmembrane pressure and cross flow velocity at an electric field of 400 V/m and pulse ratio of 3:1

	E_{punp} (kWh/m^3)				$E_{electrical}$ (kWh/m^3)			
u_0 (m/s) →	0.09	0.12	0.15	0.18	0.09	0.12	0.15	0.18
ΔP (kPa)	↓							
220	50.43	63.70	75.64	83.80	1.34	1.27	1.20	1.11
360	76.83	95.20	112.5	127.32	1.25	1.16	1.10	1.03
500	103.15	128.90	149.5	171.92	1.20	1.13	1.05	1.0
635	126.78	158.80	181.95	212.44	1.16	1.01	1.0	0.98

2.4.2.3. Analysis of Steady State Flux of Mosambi (Citrus Sinensis (L.) Osbeck) Juice

It can be observed from the figure that the energy consumption per unit volume of permeate due to applied d.c. electric field decreases with increase in both cross flow velocity and transmembrane pressure. This can be explained by the fact that increase in both cross flow velocity and transmembrane pressure lead to an increase in permeate flux. Therefore, for a fixed electric field and pulse ratio, energy consumption per unit volume of permeate decreases with increase in both cross flow velocity and transmembrane pressure. For example, at a cross flow velocity of 0.12 m/s, with increase in transmembrane pressure from 220 kPa to 635 kPa, energy consumption due to d.c. electric field decreases from 1.27 kWh /m^3 to 1.01 kWh /m^3 whereas, at a transmembrane pressure of 360 kPa, with increase in cross flow velocity from 0.09 m/s to 0.18 m/s, it decreases from 1.25 kWh /m^3 to 1.03 kWh /m^3 of permeate.

Theoretical Aspect

The expression for steady state permeate flux in case of gel layer controlled electric field assisted ultrafiltration is obtained using film theory as,

$$v_w = v_e + k \ln \frac{c_g}{c_0} \tag{2.47}$$

where, k is the mass transfer coefficient, defined as $\frac{D}{\delta_c}$; D is diffusion coefficient of the gel forming solute; and v_e is the electrophoretic velocity which can be calculated from Eqs. (2.8) and (2.9). The mass transfer coefficient under laminar flow condition can be expressed as:

$$Sh = \frac{k\, d_e}{D} = 2.1\left(ReSc\frac{d_e}{L}\right)^{\frac{1}{3}} \tag{2.48}$$

where, d_e is the equivalent diameter of the flow channel. For a thin channel, the value of d_e is 4h, where h is the half height of the channel. In mosambi juice, apart from major constituents like pectin and sugar, other components are also present such as proteins, microorganisms, acids, salts, aroma and flavor compounds. Since the diffusivity and gel concentration of actual mosambi juice differ from that of pure pectin, the estimation of permeate flux based on the values of diffusivity and gel concentration of pure pectin will be inaccurate. Moreover, the variation of viscosity with concentration has to be incorporated for better prediction. Therefore, instead of using bulk viscosity in Eq. 2.9, an effective viscosity has been suggested. An optimization method has been employed to determine which uses an initial guess for the values of these three parameters in order to minimize the following error function and obtain the correct values of the three parameters:

$$S = \sum_{i=1}^{N} \left(\frac{V_w^{exp} - V_w^{cal}}{V_w^{exp}} \right)^2 \tag{2.49}$$

BCPOL of IMSL Math library using direct complex search algorithm has been used for optimization. As a first guess, the values of diffusivity, gel concentration and effective viscosity are assumed to be the same as those for pure pectin solution. Knowing these parameters, values of steady state permeate flux are predicted. The results obtained are discussed in the subsequent sections.

The values of effective diffusivity, gel concentration and effective viscosity are estimated by optimizing the experimental values of average permeate flux using the optimizer BCPOL of IMSL Math library and the permeate flux is calculated using Eq. (2.47). The values of effective diffusivity, gel concentration and effective viscosity estimated thus are 6.05×10^{-11} m^2/s, 48.5 kg/m^3 and 7.03×10^{-3} Pa s respectively. The diffusivity is of the same order of magnitude as that obtained for pure pectin from the well known Stokes-Einstein relation (Table 2.1). The value of gel concentration is slightly higher than that reported for pure pectin (38 kg/m^3) by Pritchard et al. [25]. This indicates the presence of other gel forming material apart from pectin in mosambi juice.

Effect of Constant Electric Field on Permeate Flux

Figure 2.32 shows the variation of the average steady state average permeate flux with electric field for various transmembrane pressures at a cross flow velocity of 0.12 m/s during clarification of mosambi juice. It is evident from the figure that permeate flux increases with the electric field. This is due to electrophoretic movement of negatively charged pectin molecules away from the membrane surface (i.e. towards the positive electrode).This reduces the deposition of pectin molecules on the membrane surface and prevents the formation of a gel type layer formation over the membrane surface which leads to an enhancement of the permeate flux. For example, at a transmembrane pressure of 360 kPa and cross flow velocity of 0.12 m/s, the application of 400 V/m results in an increase in permeate flux from 8.65 L/m^2 h (in the absence of electric field) to 11.75 L/m^2 h. Furthermore, this figure also indicates that for a fixed electric field, an increase in transmembrane pressure leads to an increase in permeate flux due to enhanced driving force.

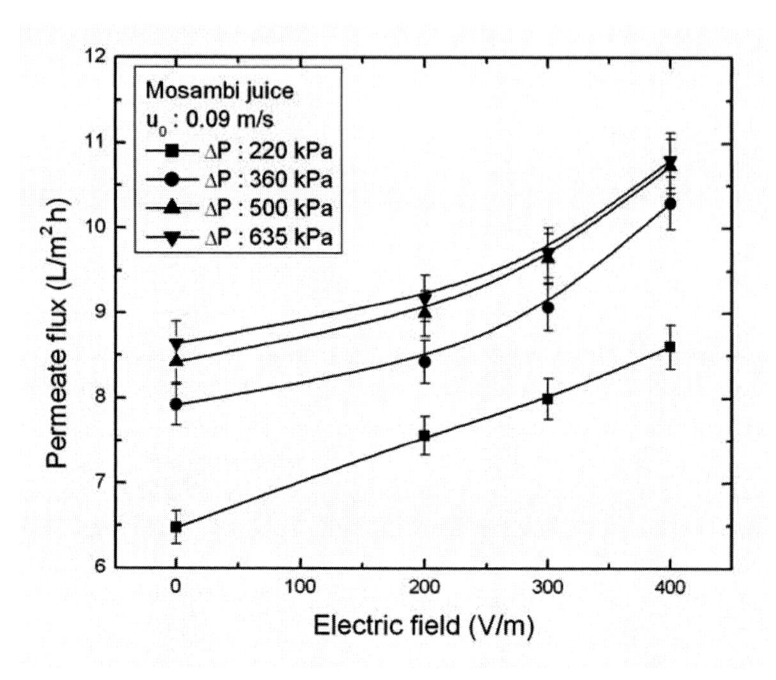

Figure 2.32. Variation of steady state average permeate flux with electric field for various transmembrane pressure during clarification of mosambi juice using 50 kDa membrane.

Effect of Cross Flow Velocity on Permeate Flux

It has been observed from Figure 2.32 that at the pressure of 360 kPa, with increase in electric field to 400 V/m the permeate flux increases by about 36% (which is an upper limit under the operating conditions employed). At other operating pressures such as 220 kPa, 500 kPa and 635 kPa, with the same increase in electric field, the flux enhancements are 32.5%, 30% and 32%, respectively. Hence the pressure 360 kPa has been selected to discuss the effect of cross flow velocity on permeate flux at different electric field strengths as shown in Figure 2.33. With increase in cross flow velocity, the sweeping effect on the gel layer over the membrane surface increases. This restricts solute particles from being deposited on the membrane surface; further it facilitates backward diffusion of the solutes from the surface to the bulk. Hence, with increase in cross flow velocity, the deposit thickness, and consequently resistance to solvent flow, decreases thereby augmenting the permeate flux.

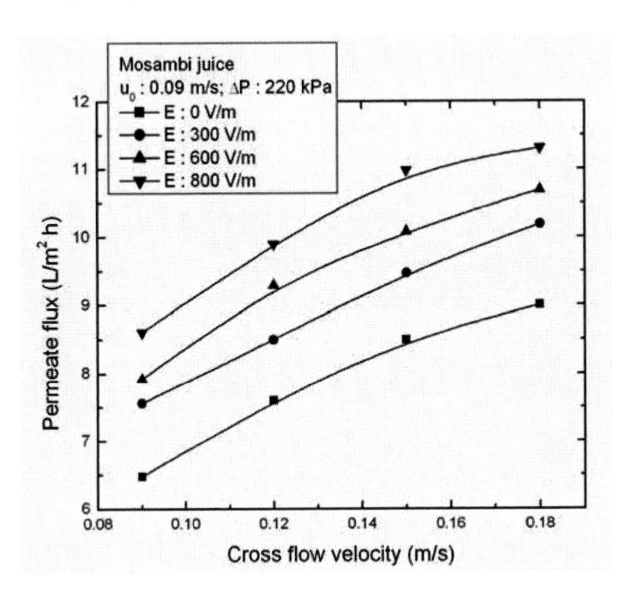

Figure 2.33. Variation of steady state average permeate flux with cross flow velocity during clarification of mosambi juice using 50 kDa membrane.

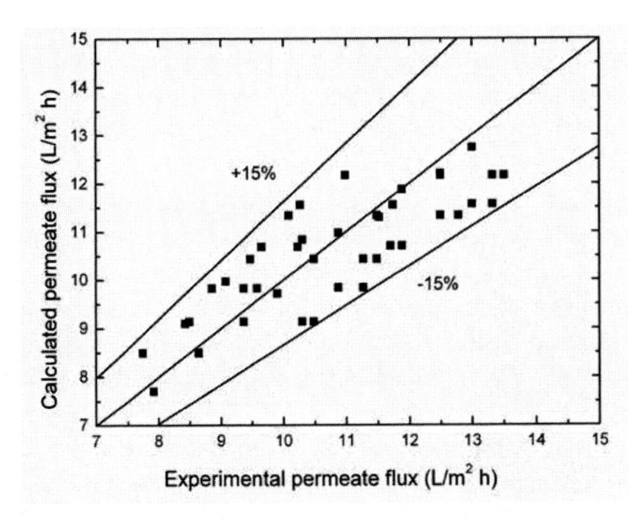

Figure 2.34. Calculated steady state permeate flux using optimized values of effective diffusivity, gel concentration and effective viscosity obtained from the optimization of cross flow electro-ultrafiltration experimental data of mosambi juice.

The effect of cross flow velocity on the enhancement in permeate flux is slightly more in the absence of electric field than at 400 V/m. At higher electric field, the layer deposited on the membrane surface is very thin due to electrophoresis. For example, at 400 V/m, with increase in cross flow velocity from 0.09 m/s to 0.18 m/s, the permeate flux increase by about 25% whereas, this augmentation is about 27% in the absence of electric field under otherwise identical operating condition.

Figure 2.34 shows the comparison between the experimental and estimated values of average permeate flux under various conditions of cross-flow velocity, electric field and transmembrane pressure. From the figure it is clear that the estimated permeate flux values are within ±15% to the experimental values.

Effect of Pulsed Electric Field

Effect of Pulse Ratio

Figure 2.35 and 2.36 show the variation in steady state average permeate flux with pulse (on time/off time) ratio for various electric field at a transmembrane pressure of 360 kPa and a cross flow velocity of 0.12 m/s during clarification of mosambi juice respectively. The symbols are for the experimental permeate flux, whereas, the solid lines are model predicted permeate flux and the dashed lines are estimated gel layer thickness. As can be seen from the figure, the predictions are very close to the experimental values of permeate flux. In fact it can be seen that the model predictions are within ±5% of the experimental results. The off-time is always fixed at 1s. It can clearly be seen that an increase in electric field results in a decrease in the estimated gel layer thickness (less resistance to flow) and associated increase in flux as observed in case of synthetic juice. As discussed earlier with increase in electric field, movement of charged pectin molecules away from the membrane surface increases leading to restriction of the formation of gel-type layer over the membrane surface. Hence permeate flux increases. For example, at a transmembrane pressure of 360 kPa, cross flow velocity of 0.12 m/s, pulse ratio of 3:1 and electric field of 400 V/m, the permeate flux increases from 8.65 L/m^2 h to 11.2 L/m^2 h compared to zero electric field and corresponding values of gel layer thickness decreases from 5.4 μm to 4.1 μm (Figure 2.36). Experimental results confirm that for a given set of operating conditions, 94-97.5% of the maximum permeate flux (electric field always on) is obtained at a pulse ratio of 3:1 (i.e., 3 s on followed by 1 s off time). Further increase in pulse ratio does not improve the permeate flux significantly and similar trends are observed for all electric fields. These results are consistent with the results of synthetic juice.

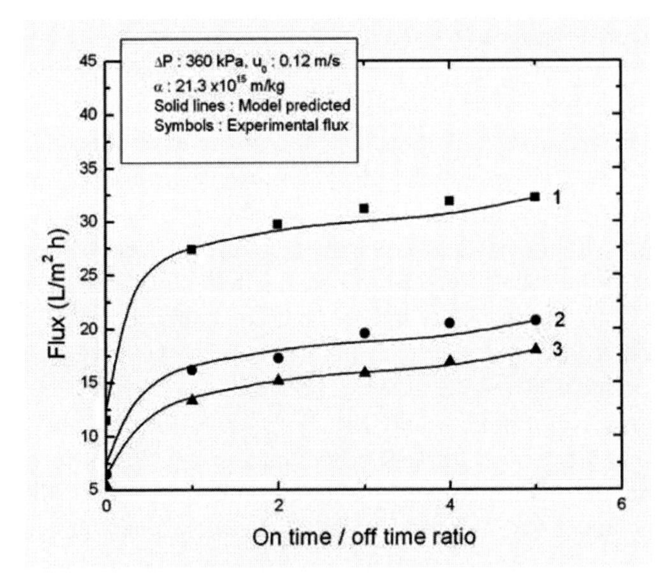

Figure 2.35. Variation in steady state average permeate flux with pulse ratio for various electric field during clarification of mosambi juice. ■, Curve 1 for E = 400 V/m; ●, Curve 2 for E = 300 V/m; ▲, Curve 3 for E = 200 V/m.

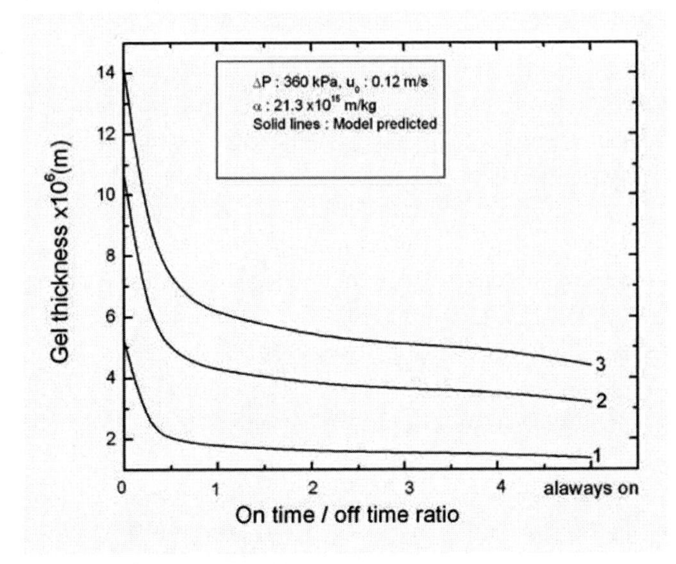

Figure 2.36. Variation in steady state gel layer thickness with pulse ratio for various electric field during clarification of mosambi juice. Curve 1 for E = 400 V/m; Curve 2 for E = 300 V/m; Curve 3 for E = 200 V/m.

Effect of Cross Flow Velocity

The variations of permeate flux and gel layer thickness with cross flow velocity for different electric fields at a pulse ratio of 3:1 are presented in Figures 2.37 and 2.38 respectively. The symbols are for the experimental permeate flux, whereas, the solid lines are model predicted permeate flux and the dashed lines are estimated gel layer thickness. As can be seen from the figure that the model predictions are within ± 3% of the experimental results. From the results shown in this figure, it is clear that for a fixed electric field, with increase in cross flow velocity permeate flux increases and gel layer thickness decreases as observed in case of synthetic juice. At higher cross flow velocity, gel layer thickness on the membrane surface is lower due to forced convection. This is a direct consequence of enhancement of turbulence in the flow channel. Furthermore, the back diffusion from membrane surface to the bulk increases as the concentration gradient from the membrane surface to the bulk increases. This leads to an increase of permeate flux at higher cross flow velocity. For example, at 400 V/m, with an increase of cross flow velocity from 0.09 m/s to 0.18 m/s permeate flux increases from 10.45 L /m^2 h to 12.5 L /m^2 h (Figure 2.38) with decrease in gel layer thickness from 4.4 μm to 3.65 μm (Figure 2.38).

The variations in steady state average permeate flux with pulse ratio for various electric fields at a transmembrane pressure of 360 kPa and cross flow velocity of 0.12 m/s during clarification of mosambi juice are shown in Figure 2.39. It may be observed from the figure that permeate flux very close to the maximum value (always on) is obtained at a pulse ratio of 3:1 as observed in case of synthetic juice (Figure 2.27). As indicated earlier, permeate flux increases with increase in electric field due to enhanced electrophoretic movement of pectin molecules away from the membrane surface (i.e., towards the positive electrode) which reduces the gel type layer formation over the membrane surface. For example, at a transmembrane pressure of 360 kPa, cross flow velocity of 0.12 m/s and pulse ratio of 3:1, by applying an electric field of 500 V/m, permeate flux is improved about 39% compared to zero electric field (0 V/m).

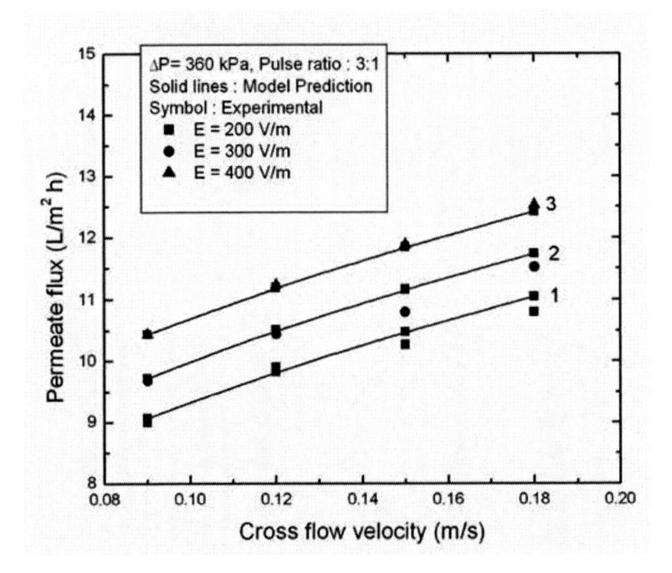

Figure 2.37. Variation in steady state average permeate flux with cross flow velocity at a pulse ratio of 3:1 for various electric field during clarification of mosambi juice. Solid lines and dashed lines are for flux and gel thickness respectively. ■, Curve 1 for E = 200 V/m; ●, Curve 2 for E = 300 V/m; ▲, Curve 3 for E = 400 V/m.

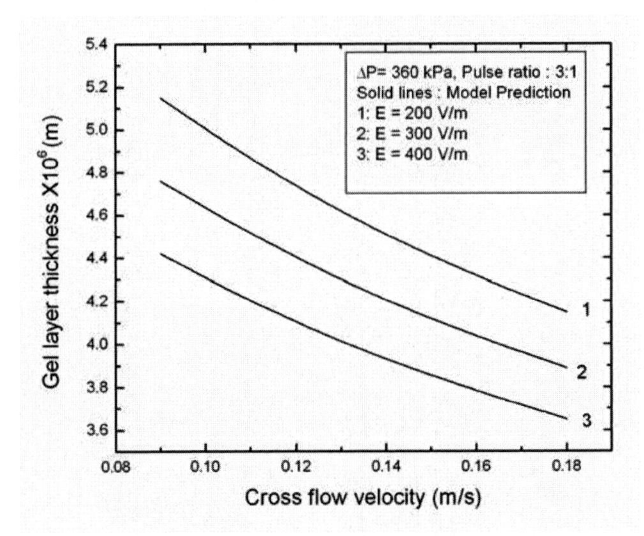

Figure 2.38. Variation in steady state gel layer thickness with cross flow velocity at a pulse ratio of 3:1 for various electric field during clarification of mosambi juice. Curve 1 for E = 200 V/m; Curve 2 for E = 300 V/m; Curve 3 for E = 400 V/m.

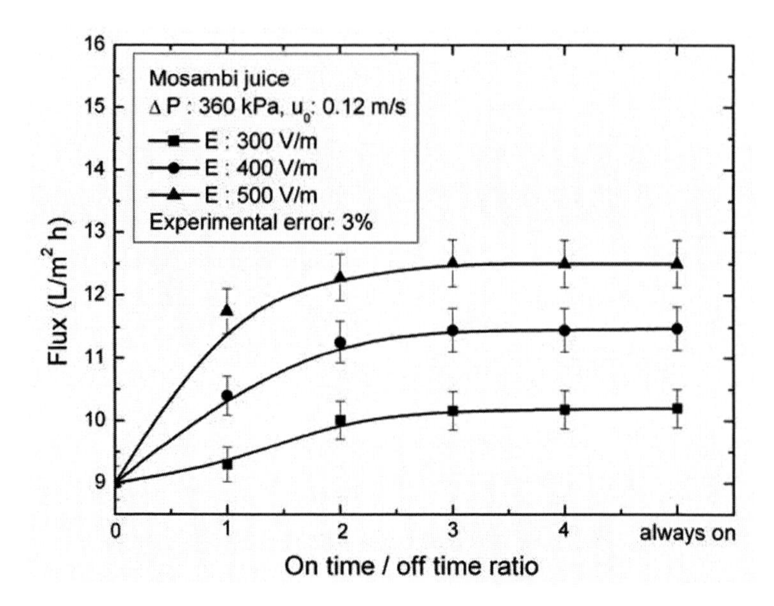

Figure 2.39.Variation in steady state average permeate flux with pulse ratio for various electric field during clarification of mosambi juice using 30 kDa membrane. The solid lines are only guides for the reader.

Table 2.11. Physico-chemical properties of permeate after clarification of mosambi juice at a transmembrane pressure of 360 kPa and a pulse ratio of 3:1

Electric field (V/m)	Cross-flow velocity (m/s)	pH	Color (A_{420})	Clarity % (T_{660})	Sucrose (^{0}Brix)	Conduc-tivity x10^4(S/m)	Viscosity $\times 10^3$ (Pa s)	Pectin (kg/m^3)	Acidity as Citric acid (kg/m^3)
0	0.09	3.97	0.22	98.4	8.4	4.0	0.87	Nil	4.0
	0.12	3.97	0.21	98.5	8.4	4.0	0.87	Nil	4.0
	0.15	3.98	0.21	98.4	8.4	4.0	0.87	Nil	4.0
	0.18	3.98	0.21	98.6	8.4	4.0	0.87	Nil	4.0
300	0.09	4.30	0.22	96.9	8.4	3.82	0.87	Nil	3.7
	0.12	4.15	0.22	97.8	8.3	3.84	0.87	Nil	3.9
	0.15	4.05	0.21	98.0	8.4	3.90	0.87	Nil	4.0
	0.18	4.00	0.21	97.8	8.3	3.95	0.87	Nil	4.0
500	0.09	4.60	0.22	97.3	8.4	3.86	0.87	Nil	3.5
	0.12	4.40	0.22	98.0	8.4	3.90	0.87	Nil	3.6
	0.15	4.36	0.21	98.5	8.4	3.90	0.87	Nil	3.6
	0.18	4.25	0.22	98.0	8.4	3.95	0.87	Nil	3.7

Characterization of Clarified Mosambi Juice

The physico-chemical properties of permeate after clarification of mosambi juice at a transmembrane pressure of 360 kPa and a pulse ratio of 3:1 are tabulated in Table 2.11.

From the results it is clear that density, conductivity, sugar content of both the feed and permeate are practically the same while viscosity is reduced from 2.45×10^{-3} Pa s to 0.87×10^{-3} Pa s due to enzyme treatment and pulsed electro-ultrafiltration. After enzyme treatment color (A_{420}), clarity (T_{660}) and acidity (i.e., as citric acid) of mosambi juice are improved from 1.25 to 1.1, 29.8% to 34% and 0.7 to 0.5 weight % respectively. There is a significant improvement of color (A_{420}) and clarity (T_{660}) after pulsed electro-ultrafiltration and no pectin is found in clarified juice. Color (A_{420}) is decreased from 1.1 to 0.22 and clarity (T_{660}) is improved from 34% to about 98% after pulsed electro-ultrafiltration. A decrease in acidity and consequently an increase in pH of permeate is observed after electro-ultrafiltration. It is further noticed that for a given electric field, with increase in cross flow velocity, pH of the permeate decreases. At higher cross flow velocity, produced OH^- ions by electrode reaction are sheared off from the electrode-membrane interface leading to a decrease in pH of the permeate. For example, with increase in cross flow velocity from 0.09 m/s to 0.18 m/s, pH of the permeate decreases from 4.6 to 4.25.

Power Consumption and pH Variation

The variation of electric power consumption and pH values of the permeate at various electric field and pulse ratio during clarification of mosambi juice at a transmembrane pressure of 360 kPa and cross flow velocity of 0.12 m/s are shown in Figures 2.40 and 2.41 respectively. Experimental result shows that for a fixed electric field of 400 V/m, pH of the permeate increases from 3.97 to 4.6 compared to without electric field, whereas, with the same increment of electric field but with a pulse ratio of 3:1, pH value increases to 4.4. This implies the advantages of a pulsed electric field.

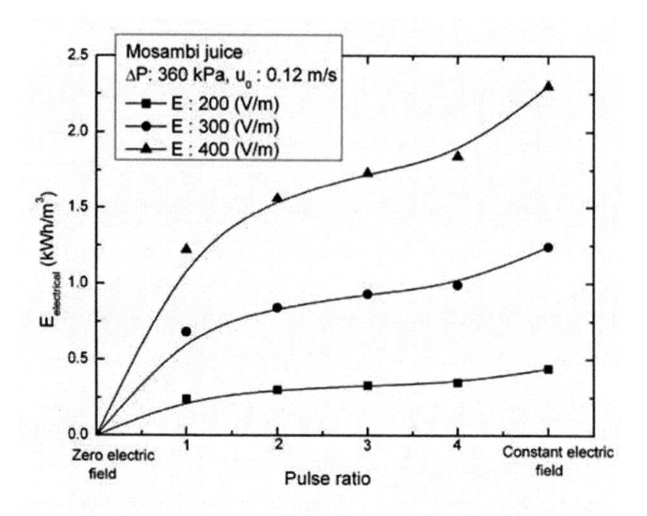

Figure 2.40. Variation of electric power consumption per unit volume of permeate with d.c. electric field at various pulse ratios at a transmembrane pressure of 360 kPa and cross flow velocity of 0.12 m/s during clarification of mosambi juice.

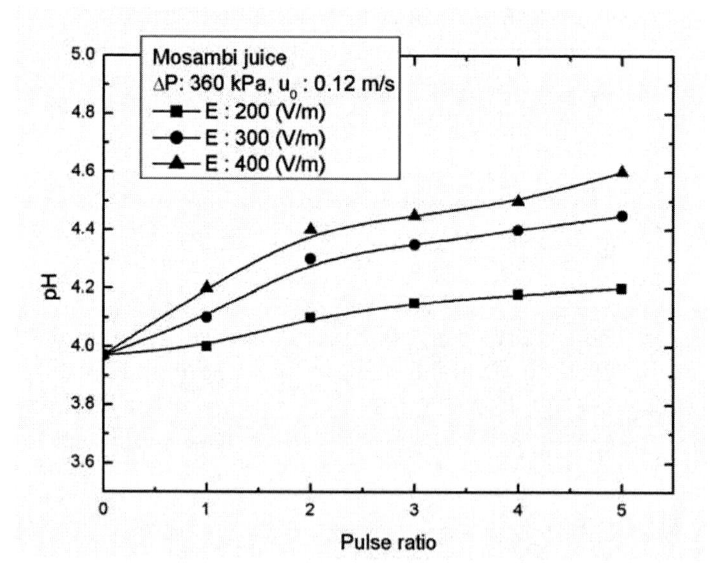

Figure 2.41. Variation of pH of the permeate with d.c. electric field at various pulse ratios at a transmembrane pressure of 360 kPa and cross flow velocity of 0.12 m/s during clarification of mosambi juice.

2.5. CONCLUSION

The effects of varying electric field during gel controlled ultrafiltration of pectin-sucrose solution are explored in this Chapter. The introduction of an electric field of appropriate polarity substantially improves the permeate flux. The trends are found to be in agreement with the physical understanding of the system. For example, the flux increases with electric field, pressure, cross-flow velocity but decreases with and concentration of pectin in solution. A theoretical approach for the prediction of permeate flux of an electric field enhanced gel controlled cross flow ultrafiltration is proposed in this work. An integral method with appropriate concentration profile in the boundary layer is used for the development of the model. The solutions are obtained for rectangular channel under laminar flow condition. It can also be observed that local permeate flux decreases sharply near the entrance region and then gradually for rest of the channel. The predictions from the proposed model are successfully compared with the experimental results under a wide range of operating conditions. The growth of the gel-type layer has been modeled taking into account the electrophoretic mobility of the charged pectin molecules in an electric field. A resistance-in-series model is proposed considering the hydrodynamic resistance and a resistance due to the growing gel layer. The model equations are solved numerically and the specific gel layer resistances are evaluated at different operating conditions. The thicknesses of the deposited gel-layer have been accurately measured by capturing the images of the deposition over the membrane surfaces using high-resolution video-microscopy and subsequent image analysis. The measured thicknesses under different operating conditions are successfully compared with the model predictions.

The drawbacks of constant electric field ultrafiltration such as high energy consumption, electrode reaction, increase in temperature, limitation to the use of feeds of high conductivity and heat sensitivity etc. can be significantly addressed using pulsed electric field. It can be seen that release of electrolytic gases with the permeate stream is more pronounced at constant electric field associated with higher pH values of the permeate. Apart from flux enhancement, use of pulsed electric field requires less energy and may lead to increase in the membrane service life due to less fouling, less frequent replacement of membrane during operation. For example, for a feed (Pectin: 3 kg/m^3, sucrose: 12 Brix) mixture, at a transmembrane pressure of 360 kPa and a cross flow velocity of 0.12 m/s, by applying constant electric field of 1000 V/m, permeate flux increases from 6.5 L/m^2 h to 25.2 L/m^2 h compared to zero electric field and associated extra electric power consumption is 0.83 kWh/m^3 of permeate. To yield almost same magnitude of permeate flux, pulsed electro-ultrafiltration with pulse ratio of 3:1 needs 0.65 kWh/m^3 of permeate. For mosambi juice, at a transmembrane pressure of 360 kPa, cross flow velocity of 0.12 m/s and pulse ratio of 3:1, by applying an electric field of 500 V/m, permeate flux is improved about 39% compared to zero electric field. Moreover, significant improvement of color and clarity are observed and no pectin is found in the clarified juice.

This study also highlights the potential utility of using external d.c. electric field to increase the productivity (throughput) during clarification of mosambi juice. For example, at a cross-flow velocity of 0.12 m/s and a transmembrane pressure of 360 kPa the permeate flux increases from 8.65 L/m^2 h to 11.75 L/m^2 h (i.e., 35.8 % flux enhancement) by applying electric field of 400 V/m, and the additional electrical power consumption is 1.16 kWh/m^3 of permeate. A comparison between experimental values of permeate flux and a model developed earlier show that the gel concentration is about 48.5 kg/m^3 and the effective diffusivity of the solute in the juice is about $6x10^{-11}$ m^2/s. The effective viscosity in the concentration boundary layer adjacent to gel layer is $7.03x10^{-3}$ Pa s which is about 7 times the bulk viscosity.

According to above findings, the electro-ultrafiltration appears to be a promising alternative for clarification of citrus fruit juice and pectin containing solutions in food processing industries in general. Moreover, the models developed in this study to quantify the effects of electric field on permeate flux should be of immense help for designing electro-ultrafiltration unit.

REFERENCES

[1] S. Alvarez, F.A. Riera, R. Alvarez, J. Coca, F.P. cuperus, S. Th Bouwer, G. Boswinkel, R.W. van Gemert, J.W. Veldsink, L.Giorno, L. Donato, S. Todisco, E. Drioli, J. Olsson, G. tragardh, S.N. Gaeta, L. Panyor, a new integrated membrane process for producing clarified apple juice and apple juice aroma concentrate, *J. Food Eng.* 46 (2000) 109-125.

[2] Cassano, B. Jiao, E. Drioli, Production of concentrated kiwifruit juice by integrated membrane process, *Food Research International* 37 (2004) 139–148

[3] E. Maccarona, S. Campisi, M.C.C, B. Fallico, C.N. Asmundo, Thermal treatments effect on the red orange juice constituents, *Industria Bevande,* 25 (1996) 335-341.

[4] Girard, L. R. Fukumoto, Membrane processing of fruit juices and beverages: a review. *Critical Reviews in Food Science and Nutrition* 40(2) (2000) 91-157.

[5] R. Thakur, R.K. Singh, A.K. Handa, Chemistry and Uses of Pectin- A Review, *Critical reviews in Food Science and Nutrition* 37 (1) (1997) 47-73.

[6] M.Z. Sulaiman, N.M. Sulaiman, M. Shamel, Ultrafiltration studies on solutions of pectin, glucose and their mixture in a pilot scale cross flow membrane unit, *Chem. Eng. J.* 84 (2001) 557-563.

[7] S. Lee, M. Elimelech, Salt cleaning of organic-fouled reverse osmosis membranes, *Water Res.* 41 (2007) 1134-1142.

[8] P. Rai, G.C. Majumdar, S DasGupta, S. De, Modeling of permeate flux of synthetic fruit juice and mosambi juice (Citrus sinensis (L.) Osbeck) in stirred continuous ultrafiltration, *LWT* 40 (2007) 1765-1773.

[9] P. Rai, C. Rai, G.C. Majumdar, S DasGupta, S. De, Resistance in series model for ultrafiltration of mosambi (Citrus sinensis (L.) Osbeck) juice in a stirred continuous mode, *J. Membr. Sci.* 283 (2006) 116-122.

[10] W.R. Bowen, H.A.M. Subuni, Electrically enhanced membrane filtration at low cross-flow velocities, *Ind. Eng. Res.* 30 (1991) 1573-1579.

[11] W.R. Bowen, H.A.M. Subuni, Pulsed electrokinetic cleaning of cellulose nitrate microfiltration membrane, *Ind. Eng. Res.* 31 (1992) 515-523.

[12] W.R. Bowen, R.S. Kingdon, H.A.M. Subuni, Electrically enhanced separation processes: the basis of in situ intermittent electrolytic membrane cleaning (IIEMC) and in situ electrolytic membrane restoration (IEMR), *J. Membr. Sci.* 40 (1989) 219-229.

[13] C.W. Robinson, M.H. Siegel, A. Condemine, C. Fee, T.Z. Fahidy, B.R. Glick, Pulsed-electric-field cross flow ultrafiltration of bovine serum albumin, *J. Membr. Sci.* 80 (1993) 209-220.

[14] H.R. Rabie, A.S. Majumdar, M.E. Weber, Interrupted electroosmosis dewatering of clay suspensions, *Sep. Technol.* 4(1) (1994) 38-46.

[15] A.K. Vijh, Electrochemical aspects of electroosmotic dewatering of clay suspension, *Drying Technol.* 13 (1-2) (1995) 215-224.

[16] S. De, P.K. Bhattacharya, Modeling of ultrafiltration process for a two-component aqueous solution of low and high (gel forming) molecular weight solutes, *J. Membr. Sci.* 136 (1999) 57-69.

[17] V.S. Minnikanti, S. DasGupta, S.De, Prediction of mass transfer coefficient with suction for turbulent flow in cross flow ultrafiltration, *J. Membr. Sci.* 157 (1999) 227-239.

[18] S. De, S. Bhattarcharya, P.K. Bhattarcharya, A. Sharma, Generalized integral & similarly solution of the concentration profile for osmotic pressure controlled ultrafiltration, *J. Membr. Sci.* 130 (1997) 99-121.
[19] J.S. Shen, R.F. Probstein, On the prediction of limiting flux in laminar ultrafiltration of macromolecular solutes, *Ind. Eng. Chem. Fundam.* 16 (1977) 459-465.
[20] R.F. Probstein, J.S. Shen, W.F. Leung, Ultrafiltration of macromolecular solution at high polarization in laminar channel flow, *Desalination* 24 (1978) 1-16.
[21] M. Cheryan, *Ultrafiltration and Microfiltration Handbook,* Technomic Publishing Company Inc., USA, 1998.
[22] F.L. Hart, H.J. Fisher, *Modern Food Analysis,* Springer, Berlin, 1971.
[23] R.J. Hunter, *Zeta potential in colloid Science: Principles and Applications,* Academic Press, London, 1981.
[24] L. Stryer, *Biochemistry,* Freeman, New York, 1998.
[25] M. Pritchard, J.A. Howell, R.W. Field, The ultrafiltration of viscous fluids, *J. Membr. Sci.* 102 (1995) 223-235.
[26] S. Bhattacharjee, A. Sharma, P.K. Bhattacharya, A unified model for flux prediction during batch cell ultrafiltration, *J. Membr. Sci.* 111 (1996) 243-258.
[27] S. Pal, A. Ghosh, T.B. Ghosh, S. De, S. DasGupta, Optical quantification of fouling during nanofiltration of dyes, *Sep. Purif. Tech.* 52 (2006) 372-379.
[28] B. Sarkar, S. Pal, T.B. Ghosh, S. De, S. DasGupta, A study of electric field enhanced ultrafiltration of synthetic fruit juice and optical quantification of gel deposition, *J. Membr. Sci.* 311 (2008) 112–120.
[29] R.B. Bird, W.E. Stewart, E.N. Lightfoot, *Transport phenomena,* John Wiley, Singapore, 2002.
[30] S. De, S. Bhattarcharya, P.K. Bhattarcharya, A. Sharma, Generalized integral & similarly solution of the concentration profile for osmotic pressure controlled ultrafiltration, *J. Membr. Sci.* 130 (1997) 99-121.
[31] S.M. Oussedik, D. Belhocine, H. Grib, H. Lounici, D.L. Piron, N. Mameri, Enhanced ultrafiltration of bovine serum albumin with pulsed electric field and fluidized activated alumina, *Desalination* 127 (2000) 59-68.

Chapter 3

SEPARATION AND FRACTIONATION OF PROTEIN SOLUTION

ABSTRACT

This Chapter is divided into two sections. In the first section, application of external d.c. electric field for the improvement of permeate flux during ultrafiltration of single protein solution is discussed. In this section, a theoretical model is developed to predict permeate flux under laminar flow regime including the effects of external d.c. electric field and suction through the membrane for osmotic pressure governed ultrafiltration of single protein solution. The governing equations of the concentration profile in the developing mass transfer boundary layer in a rectangular flow channel are solved using similarity solution technique. Effect of d.c. electric field on the variation of membrane surface concentration and permeate flux along the length of the flow channel is quantified using this model. An expression of Sherwood number relation for estimation of mass transfer coefficient is also derived. The analysis revealed that there is a significant effect of electric field on the mass transfer coefficient. In the second part, the effect of d.c. electric field on the permeate flux and retention characteristics during fractionation of aqueous solution of two proteins has been discussed. In this section, a theoretical model based on film theory is developed to predict permeate flux and observed retention under laminar flow regime including the effects of external d.c. electric field for osmotic pressure governed ultrafiltration of aqueous solution of two proteins. The governing equations of the concentration profile in the developed mass transfer boundary layer in a rectangular channel are solved numerically. The model parameters such as real retention and protein-protein interaction parameter are evaluated by optimizing the experimental values of steady state permeate flux and permeate concentration. The experimental results reveal that there is a significant effect of electric field on both permeate flux and observed retention.

NOMENCLATURE

A, A_1	Parameters in Eqs.(3.11, A.8)
a	Effective hard sphere radius of BSA in solution, m
$a_{1,2,3,4}$	Coefficients in Eq.(3.10 and A.14)
A_H	Hamaker constant, J
A_h	Effective area of a single particle at a hypothetical plane, m^2
B	Constant in Eq.(3.19)
$b_{1,2,3,4}$	Coefficients in Eq.(A.15)
c	Solute concentration, kg/m^3
c_0	Solute concentration at bulk, kg/m^3
c_p	Permeate concentration, kg/m^3
c_m	Membrane surface concentration, kg/m^3
c_m^*	Dimensionless membrane surface concentration (c_m/c_o)
c*	Dimensionless concentration (c/c_0)
C_1	Ionic strength, $kmol/m^3$
d_e	Equivalent diameter of channel, m
d	Distance to surface of shear, m
δ*	Dimensionless concentration boundary layer thickness (δ/h)
D	Diffusivity of BSA, m^2/s
d_1	Interparticle distance, m
D_i	Dielectric constant, dimensionless
e	Electronic charge, 1.6×10^{-19}, C
E	Electric field, V/m
$f(d_1)$	Electrostatic interaction force, N
$F_A(d_1)$	Attractive interparticle force, N
h	Half of channel height, m
I_1	Definite integral in Eq.(3.26)
K_B	Boltzmann constant (1.38×10^{-23}), J/K

k	Mass transfer coefficient, m/s
κ^{-1}	Debye length, m
k', k''	Integration constants in Eqs.(3.24,3.25)
L	Channel length, m
Lp	Membrane permeability, m/Pa s
M_w	Molecular weight, gm/gm mole
N_A	Avogadro's number, mol^{-1}
n	Number concentration of electrolyte in the bulk, m^{-3}
ΔP	Transmembrane pressure, Pa
$\overline{Pe_w}$	Dimensionless length average permeate flux
Pe_w	Dimensionless permeate flux
Pe_e	Dimensionless permeate flux due to electric field
r_{cell}	Radius of spherical cell, m
Re	Reynolds number at the bulk condition, dimensionless
Sc	Schmidt number at the bulk condition, dimensionless
Sh	Local Sherwood number, dimensionless
$\overline{Sh}$	Average Sherwood number, dimensionless
S_{cell}	Surface area of spherical cell, m^2
T	Absolute temperature, K
u_e	Electrophoretic mobility, ms^{-1}/Vm^{-1}
u	Axial velocity, m/s
u_o	Average bulk velocity, m/s
v_e	Electrophoretic velocity, m/s
V_A	Attractive interaction energy, J
v_w	Permeate flux, m^3/m^2 s
$\overline{v_w}$	Average permeate flux, m^3/m^2 s
v	Velocity component in normal direction, m/s
x	Axial distance, m
x^*	Dimensionless axial distance
X	Parameter in Eq.(A.12)
y	Normal distance, m
y^*	Dimensionless normal distance
z	Valency, dimensionless

Greek Letters

α	$[= \kappa (a + d)]$, dimensionless
β	$(= \kappa r_{cell})$, dimensionless
μ	Viscosity, Pa s
η	Similarity parameter
δ	Concentration boundary layer thickness, m
ξ	Zeta potential of particle, V
$\xi_{effective}$	Effective zeta potential of particle, V
ξ^1	Reduced zeta potential $[= \frac{ez\xi}{K_B T}]$, dimensionless
$\Phi_{r = r_{cell}}$	Electrostatic potential at the cell boundary, V
Ψ	Reduced electrostatic potential $[= \frac{ez\Phi}{K_B T}]$, dimensionless
ε_o	Permittivity of vacuum (8.854×10^{-12}), $CV^{-1}m^{-1}$
π	Total osmotic pressure, Pa
$\Delta\pi$	Osmotic pressure difference across the membrane, Pa
$\pi_{entropic}$	Osmotic pressure due to entropy change, Pa
$\pi_{electric}$	Osmotic pressure due to electrostatic interaction, Pa
φ	Particle volume fraction, dimensionless
φ_{cp}	Particle volume fraction at close packing, dimensionless
ρ	Bulk density, kg/m^3

Subscript

1	For BSA
2	For Lysozyme
0	Bulk condition
m	Membrane surface
p	Permeate

Superscript

* dimensionless

3.1. ELECTRIC FIELD ASSISTED ULTRAFILTRATION OF PROTEIN FROM AQUEOUS SOLUTION

Separation and purification of proteins is a crucial process in the rapidly growing bio-processing industries due to their wide range of applications in biomedical, pharmaceutical, dairy, and food industries. In recent years, membrane separation processes have attracted a substantial attention for protein purification throughout the downstream processes, largely replacing the expensive conventional technique such as chromatography, electrophoresis etc., due to their unique separation capability, low energy consumption, and high throughput while maintaining product under ambient conditions. However, the performance of ultrafiltration and microfiltration processes for protein fractionation is restricted due to concentration polarization i.e., the accumulation of rejected protein molecules near the membrane surface, and the fouling, the adsorption of protein molecules on the membrane surface. The high concentration of rejected proteins at the membrane surface results either in a high osmotic pressure difference across the membrane, which significantly reduces the effective pressure drop across the membrane, or in the formation of a deposit or gel-layer.

In an aqueous solution of proteins, there are mainly three types of colloidal interactions, namely, electrostatic repulsion of double layers, dispersion forces (London-van der Waals interaction), and entropic pressure. Electrostatic and van der Waals interactions are characterized by fixed charges and material polarizability, respectively [1]. The summation of these forces gives the basis of the DLVO theory of colloidal stability [2]. The entropic pressure is generated due to configurational entropy of the colloids in the system which is significant for smaller size colloidal particles, i.e., protein molecules. The importance of other interactions such as steric effect, hydration forces, and hydrophobic forces has been recognized [3]. These interactions have strong influence on diffusivity, viscosity, and osmotic pressure. During ultrafiltration process, the osmotic pressure controls the spatial distribution of particles in the concentration polarized layer and hence the rate of filtration [4]. Thus, the development of quantitative model, leading to prediction of permeate flux and process selectivity, is a pre-requisite for the design and successful application of any membrane separation process module. Current research and development efforts toward the improvements in selectivity with high system throughput should facilitate membrane systems to play an important role in the next generation of bio-processes. Bowen et al. [4] have developed a theory based on a Wigner-Seitz cell [5] approach considering the multiparticle electrostatic interactions, dispersion forces, hydration forces, and configurational entropy effect through an extended DLVO theory for osmotic pressure calculation. They have found the excellent agreement of this model with the experimental data for the ultrafiltration of silica colloids. Bellara and Cui [6] have used the Maxwell–Stefan equations to predict the permeate flux in ultrafiltration of BSA solutions. Vasan et al. [7] have derived an analytical expression based on the Maxwell–Stefan equation and the Gouy-Chapman model to explain the concentration profile of charged solute in the concentration polarization layer.

Application of electric field is found to be a promising alternative for large-scale separation, purification, and fractionation of protein solution. Enevoldsen et al. [8] have investigated the effect of electric field during ultrafiltration of amylase enzyme and showed the flux improvement in the order of 3-7 times. They have also reported the process design and economic aspects of electro-ultrafiltration of amylase enzyme [9]. The application of electric field is studied in a two-sided filtration device with flushed electrode for the separation of BSA and Lysozyme [10]. Zumbusch et al. [11] have shown the possibilities of using an alternating current to enhance the flux during filtration of BSA. During electro-filtration of quartz suspension, the importance of electro chemical reaction on the filtration kinetics is investigated by Saveyn et al. [12].

In this section, the application of electric field is explored in detail during the separation of BSA protein from aqueous solution. It has been observed that the transport of solutes towards the membrane is strongly influenced by the nature of hydrodynamic and concentration boundary layers. Therefore, a quantitative and precise analysis of concentration polarization is required for proper designing and optimization of membrane module. For this purpose, an accurate estimation of mass transfer coefficient is required, which is generally derived from Sherwood number correlations, obtained from heat-mass transfer analogies. In literature, various Sherwood number relations applicable to membrane system are available and extensively reviewed [13,14]. These available correlations are based on some assumption which are mentioned as: (i) they are derived for non porous conduit; hence suction effect is not considered, (ii) concentration boundary layer is assumed to be fully developed (stagnant film) over most of the channel which may not be true for ultrafiltration of macromolecules (Schmidt number more than 10000) where concentration boundary layer is developing, (iii) the variation of physical properties like, viscosity, density, diffusivity with concentration are not considered. A method of quantification of concentration profile in a thin polarized layer near the membrane surface in framework of filtration of charged solute is reported [15]. However, this work [15] also assumes a stagnant concentration boundary layer on the membrane surface which may not be applicable in long ultrafiltration channel. Hence, appropriate correlation of Sherwood number for the case of electric field enhanced ultrafiltration is not available in literature. In view of this, a generalized Sherwood number relationship for calculating the mass transfer coefficient is urgently warranted for proper design of electro-ultrafiltration system. An expression of Sherwood number for osmotic pressure controlled cross-flow ultrafiltration of BSA, assisted by external electric field in the laminar flow regime is developed here. During the derivation, the effects of suction and developing mass transfer boundary layer are also included. However, it may be mentioned here that concentration dependency on the physical properties, e.g., viscosity, diffusivity, etc. is not included in the present discussion.

3.1.1. Mass Transfer Analysis during Electric Field Assisted Ultrafiltration

Extensive studies on bovine serum albumin (BSA) show that the osmotic pressure of this protein plays an important role during its filtration. BSA is highly soluble and gel formation is unlikely to occur even at the high concentrations on the membrane surface

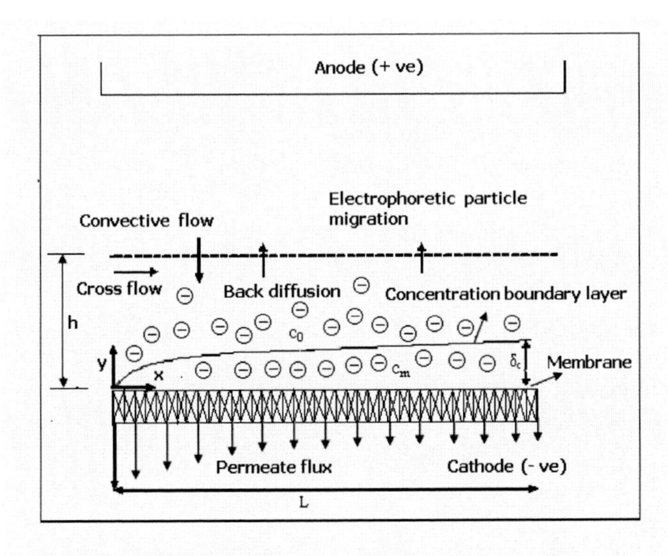

Figure 3.1. Schematic of formation of external concentration boundary layer over the membrane surface in presence of external d.c. electric field. The thickness of the boundary layer has been greatly enlarged for clarity.

Therefore, ultrafiltration of aqueous solution of BSA is assumed to be osmotic pressure controlled. The flow geometry of the electric field enhanced ultrafiltration is presented in Figure 3.1.

Following are the assumption during theoretical developments presented in this study: (i) laminar, steady state flow, (ii) velocity boundary layer is completely developed (small entrance length); (iii) velocity is small compared to the axial cross flow velocity, so that the velocity profile in the flow channel remains undistorted; (iv) no adsorption of solute over the membrane surface; (v) the channel gap is sufficiently small so that electric field is assumed to be uniform; (vi) the flow is two dimensional

The steady state solute mass balance in the rectangular channel near the membrane surface is given by

$$u\frac{\partial c_1}{\partial x} + (v + v_{e1})\frac{\partial c_1}{\partial y} = D_1 \frac{\partial^2 c_1}{\partial y^2} \tag{3.1}$$

where, v_{e1} is the electrophoretic velocity. The expression of electrophoretic velocity is presented in Eq. (2.8). The electrophoretic mobility can be expressed using Eq. (2.9).

Assuming hydrodynamic velocity profile to be fully developed, the x-component velocity becomes [16]

$$u = \frac{3}{2} u_o \left[1 - \left(\frac{y\text{-}h}{h} \right)^2 \right] \tag{3.2}$$

Within the thin concentration boundary layer, the term $\frac{y^2}{h^2}$ can be neglected and the x-component velocity profile can be simplified as,

$$u = \frac{3u_0 y}{h} \tag{3.3}$$

where, h is the half of the channel height.

Since the thickness of concentration boundary layer is extremely small as $\delta_{c1} \propto \frac{1}{Sc_1^{\frac{1}{3}}}$ and Schmidt number is quite large due to low diffusivity of BSA molecule, the y-component velocity within concentration boundary layer is approximated as [17],

$$V = - V_w \tag{3.4}$$

The initial and boundary conditions of Eq. (3.1) are,

$$\text{at } x = 0\,, \; c_1 = c_{01} \tag{3.5}$$

$$\text{at } y = \delta_{c1}\,, \; c_1 = c_{01} \tag{3.6}$$

where, δ_{c1} is the thickness of the concentration boundary layer.

As pointed out earlier, thickness of concentration boundary is extremely small for high molecular weight solute (having lower diffusivity and higher Schmidt number). Therefore, in most of the channel cross-section, the fluid particles do not experience the presence of concentration boundary layer. Thus, for the sake of mathematical simplicity the boundary condition presented by Eq. (3.6) can be rewritten as,

$$\text{at } y = \propto\,, \; c_1 = c_{01} \tag{3.7}$$

During ultrafiltration, solutes are convected by the pressure gradient towards the membrane surface. The rejected solutes tend to accumulate over the membrane surface. Since the membrane surface concentration is higher than the bulk concentration, back diffusion of rejected solutes from the surface towards the bulk of the solution takes place. In presence of an external d.c. electric field (with the top surface having opposite polarity as that of the charged particles), the particles move away from the membrane surface due to electrophoresis. At steady state, the rate of convective movement of solute particles towards the membrane surface is equal to the rate of migration of solute particles away from the membrane surface due to both back diffusion and electrophoresis (Figure 3.1). Therefore, the boundary condition at the membrane surface can be written as,

$$\text{at } y = 0, \; D_1 \frac{\partial c_1}{\partial y} + (v_w - v_{e1})(c_{m1} - c_{p1}) = 0 \tag{3.8}$$

In an osmotic pressure governed ultrafiltration, solvent flow through the porous membrane can be expressed by the phenomenological equation as,

$$v_w = L_p \; (\Delta P - \Delta\pi) \tag{3.9}$$

Where $\Delta\pi = \pi_m - \pi_p$

The osmotic pressure of the solution can be written it terms of solute concentration as,

$$\pi_1 = a_1 c_1 + a_2 c_1^2 + a_3 c_1^3 + a_4 c_1^4 \tag{3.10}$$

For fully retentive membrane, permeate concentration (c_{p1}) is taken as zero. The methodology for calculating the osmotic pressure of BSA solution at a particular pH and salt concentration is adopted from Bowen et al. [4,18,19] and is presented in the appendix.

The governing mass balance equation, Eq. (1) can be non-dimensionalized as,

$$Ay^* \frac{\partial c_1^*}{\partial x^*} - \left(\frac{Pe_w - Pe_{e1}}{4}\right)\frac{\partial c_1^*}{\partial y^*} = \frac{\partial^2 c_1^*}{\partial y^{*2}} \tag{3.11}$$

where, $x^* = \frac{x}{L}$; $c_1^* = \frac{c_1}{c_0}$; $A = \frac{3}{16}\left(ReSc_1 \frac{d_e}{L}\right)$; $Pe_{e1} = \frac{v_{e1} d_e}{D_1}$; $Pe_w = \frac{v_w d_e}{D_1}$;

$$y^* = \frac{y}{h}$$

For a thin channel the equivalent diameter is defined as [20], $d_e = 4h$. The initial and boundary conditions are non-dimemsionalized as,

$$\text{at } x^* = 0, \; c_1^* = 1 \tag{3.12}$$

at $y^* = \infty$, $c_1^* = 1$ (3.13)

$$\text{and at } y^* = 0 \; , \; \frac{\partial c_1^*}{\partial y^*} + \left(\frac{Pe_w - Pe_{e1}}{4} \right) c_1^* = 0 \tag{3.14}$$

Since the governing equation, Eq. (3.11) is a parabolic partial differential equation, and one of its boundary condition is at infinity, Eq. (3.13), a solution of Eq. (3.11) using similarity method exists.

An order of magnitude analysis of Eq. (3.11) at the edge of concentration boundary layer (δ_{c1}) results,

$$A\,\delta_{c1}^* \frac{\Delta c^*}{x^*} \approx \frac{\Delta c^*}{\delta_{c1}^{*\,2}} \tag{3.15}$$

The above equation presents an order of magnitude of thickness of concentration boundary layer

$$\delta_{c1}^* = \left(\frac{x^*}{A} \right)^{\frac{1}{3}} \tag{3.16}$$

Hence, the similarity parameter is defined as,

$$\eta = \frac{y^*}{\delta_{c1}^*} = \left(\frac{A}{x^*} \right)^{\frac{1}{3}} y^* \tag{3.17}$$

In terms of c_1^* and η, the governing partial differential equation, Eq. (3.11) becomes a second order differential equation as

$$\frac{d^2 c_1^*}{d\eta^2} = \left[-\left(\frac{Pe_w - Pe_{e1}}{4} \right) \left(\frac{x^*}{A} \right)^{\frac{1}{3}} - \frac{\eta^2}{3} \right] \frac{dc_1^*}{d\eta} \tag{3.18}$$

In absence of electric field, permeate flux is inversely proportional to the thickness of concentration boundary layer as,

$$v_w \propto \frac{1}{\delta_{c1}}$$

In order to incorporate the effect of oppositely directed electrophoretic mobility, the above relation is modified as,

$$v_w - v_{e1} \propto \frac{1}{\delta_{c1}}$$

As observed from Eq. (3.16), $\delta_{c1} \propto x^{\frac{1}{3}}$

Hence, $(v_w - v_{e1})\, x^{\frac{1}{3}} = \text{constant}$

Thus, the first term on RHS of Eq. (3.18) can be written as

$$\left(\frac{Pe_w - Pe_{e1}}{4}\right)\left(\frac{x^*}{A}\right)^{\frac{1}{3}} = B \tag{3.19}$$

where, B is a constant.

Therefore Eq. (3.18) becomes

$$\frac{d^2c_1^*}{d\eta^2} = \left(-B - \frac{\eta^2}{3}\right)\frac{dc_1^*}{d\eta} \tag{3.20}$$

The corresponding boundary conditions in non-dimensional form are

$$\text{at } \eta = \infty,\ c_1^* = 1 \tag{3.21}$$

$$\text{and at } \eta = 0,\ \frac{dc_1^*}{d\eta} + Bc_1^* = 0 \tag{3.22}$$

Eq. (3.20) can be solved along the boundary conditions, Eqs. (3.21) and (3.22) and the solution is,

$$c_1^*(\eta) = k' \int \exp\left(-\frac{\eta^3}{9} - B\eta\right) d\eta + k'' \tag{3.23}$$

The integration constants k' and k'' can be evaluated using the Eqs. (3.21) and (3.22) as,

$$k' = \frac{-B}{1 - BI_1} \tag{3.24}$$

and $$k'' = \frac{1}{1 - BI_1} \tag{3.25}$$

where, the definite integral (I_1) is given as,

$$I_1 = \int_0^\infty \exp\left(-\frac{\eta^3}{9} - B\eta\right) d\eta \tag{3.26}$$

The average permeate flux over the length of the channel can be written from Eq. (3.19) as,

$$\overline{Pe_w} = \int_0^1 Pe_w(x^*)\, dx^* = Pe_{e1} + 6\ B\ A^{\frac{1}{3}} \tag{3.27}$$

The expression of B thus obtained as,

$$B = \frac{\left(\overline{Pe}_w - Pe_{e1}\right)}{6\ A^{\frac{1}{3}}} \tag{3.28}$$

Estimation of Mass Transfer Coefficient

From solute (BSA) mass balance at the membrane surface, the mass transfer coefficient, k_1 is defined as,

$$k_1 \left(c_{m1} - c_{01}\right) = - D_1 \left(\frac{\partial c_1}{\partial y}\right)_{y=0} \qquad (3.29)$$

Eq. (3.29) can be written in terms of similarity parameter as,

$$\frac{Sh_1}{4}\left(c^*_{m1} - 1\right) = - \left(\frac{A}{x^*}\right)^{\frac{1}{3}} \left(\frac{dc_1^*}{d\eta}\right)_{\eta=0} \qquad (3.30)$$

c^*_{m1} and $\left(\frac{dc_1^*}{d\eta}\right)_{\eta=0}$ can be substituted in terms of integration constant k_1 and k_2 and Eq. (3.30) can be expressed as,

$$\frac{Sh_1}{4} = - \left(\frac{A}{x^*}\right)^{\frac{1}{3}} \frac{k'}{k'' - 1} \qquad (3.31)$$

Substituting the values of k' and k'' from Eqs. (3.24) and (3.25), Eq. (3.31) can be expressed in terms of definite integral (I_1) as,

$$Sh_1 \; (x^*) = \frac{2.3}{I_1}\left(ReSc_1 \frac{d_e}{L}\right)^{\frac{1}{3}} x^{*\frac{-1}{3}} \qquad (3.32)$$

The length average Sherwood number can be obtained by integrating Eq. (3.32),

$$\overline{Sh_1} = \int_0^1 Sh_1 \; (x^*)\, dx^* = \frac{3.45}{I_1}\left(ReSc_1 \frac{d_e}{L}\right)^{\frac{1}{3}} \qquad (3.33)$$

where, I_1 is given by Eq. (3.26).

The values of definite integral (I_1) for different values of $\left(\overline{Pe}_w - Pe_{e1}\right)$ are calculated from Eq. (3.28) and Eq. (3.26). Figure 3.2 shows the plot of definite integral (I_1) as a function of $\left(\overline{Pe}_w - Pe_{e1}\right)$ for different values of (Re Sc d_e/L).

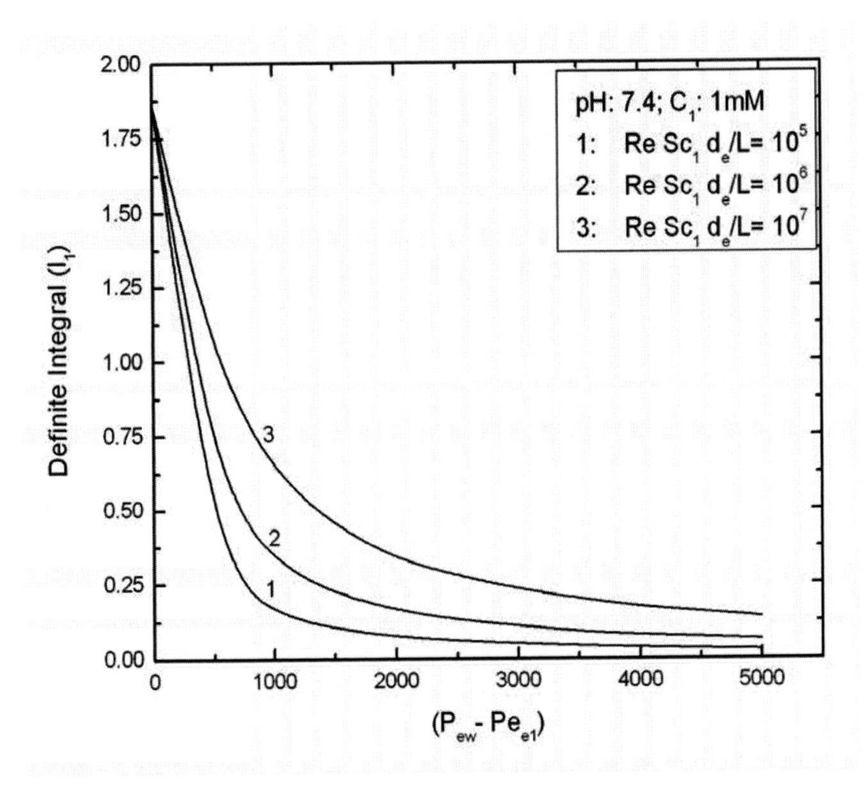

Figure 3.2. The Variation of Definite Integral (I_1) with $\left(\overline{Pe}_w - Pe_{e1}\right)$ for different values $(\text{Re Sc}_1\ d_e/ L)$.

Prediction of c_{m1} and v_w

During electric field enhanced ultrafiltration, v_w is a strong function of c_m, which is a function of applied d.c. electric field, hydrodynamic conditions etc. Therefore, for better understanding of extent of concentration polarization and nature of flux decline behavior, estimation of c_{m1} is required.

Algorithm Used for Estimation of c_m and v_w

1. For a particular x^*, a value of c_{m1}^* is guessed.
2. $\Delta\pi$ is calculated from Eq. (A.3.14).
3. v_w and Pe_w are calculated from Eq. (3.9).
4. The value of B is evaluated from Eq. (3.19).
5. I_1 is calculated from Eq. (3.26).
6. c_{m1}^* is computed from Eq. (3.25).
7. If $|c_{m1}^*(\text{calculated}) - c_{m1}^*(\text{guessed})| < 0.001$, the programme is terminated and the corresponding values of c_{m1}^* and v_w are recorded. If not another value of c_{m1}^* is guessed in step (1) and process is continued till the given convergence is achieved.
8. Then x^* is increased to $x^* + \Delta x^*$.
9. Steps 2 to 8 are repeated till $x^* = 1.0$.
10. $\overline{v_w}$ and $\overline{Pe_w}$ are calculated from Eq. (3.27).

3.1.2. Importance of Membrane Surface Charge

Electro-kinetic properties of membrane and the solute have a strong effect on the development of concentration polarization over the membrane surface [21]. During ultrafiltration of macromolecules such as polysaccharide, proteins etc, the permeate-flux and solute transmission is strongly influenced by nature and extent of solute-solute, solute-membrane interaction, which in turn depends on the electrochemical properties of membrane and solutes [22,23]. Ultrafiltration membranes are assumed to be a bundle of charged capillaries. Electro-kinetic methods such as streaming potential measurement are generally used to evaluate surface charge characteristics of different types of membranes [24-28]. Therefore, the knowledge of membrane surface charge and the appropriate control of both the permeate flux and the retention (or the transmission) of the solutes in the ultrafiltration process is of great interest. In this section surface charge characteristic of commercially available ultrafiltration membranes using streaming potential measurement technique is reported. Zeta potential of 30 kDa molecular weight cut-off (MWCO) Polyphenylene ethersulfone and 20 kDa molecular weight cut-off Polyethersulfone membrane are determined as a function of pH and ionic strength. The effect of solute-solute and solute-membrane interaction on permeate flux has been discussed during separation of protein, bovine serum albumin from aqueous solution, using 30 kDa molecular weight cut-off membrane.

Electro-kinetic Theory

When membranes come in contact with aqueous solution, it acquires surface charge due to dissociation of ionic groups present at the surface or adsorption of ions from the solution. To maintain electro neutrality of the solution, the membrane surface charge is balanced by opposite charge in the liquid medium. Due to electrostatic interaction, counter ions tend to concentrate close to the surface while co-ions move away from the surface. This leads to the formation of electrical double layer where the potential decrease exponentially from the surface to bulk of the solution.

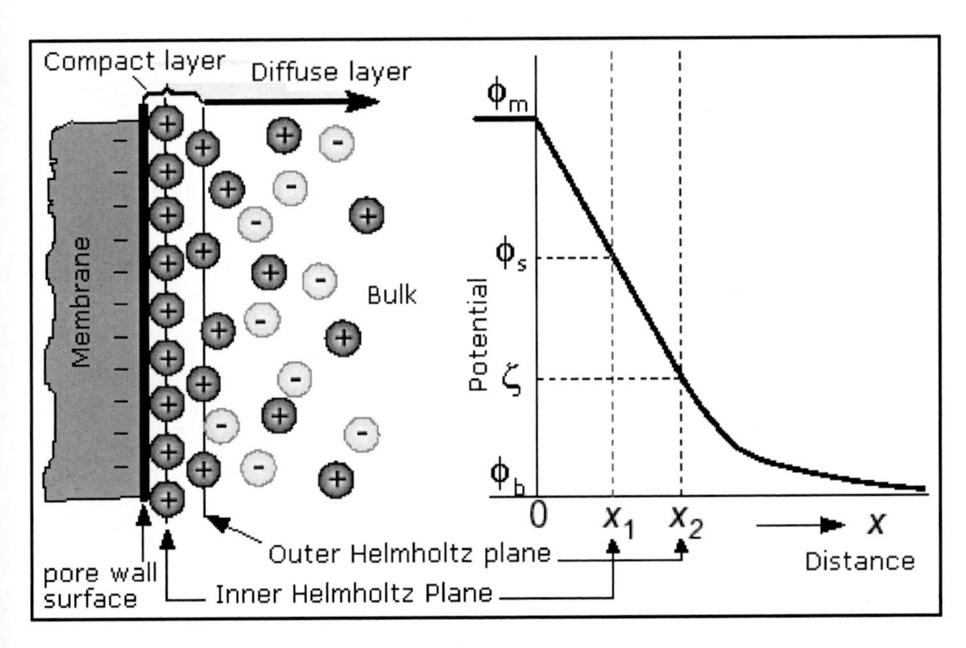

Figure 3.3. Helmholtz planes close to membrane pore wall and potentials at Helmholtz planes.

The electrical double layer is composed of compact and diffuse layer. Compact layer consists of inner Helmholtz plane and outer Helmholtz plane. The inner Helmholtz plane is assumed to be drawn through the center of absorbed and partially dissolved ions while the outer Helmholtz plane is assumed to be formed through the center of completely dissolved ions which are attached to the surface by Coulombic attraction (Figure 3.3). Gouy-Chapman model describes the charge distribution in the diffuse double layer. The potential at the plane of shear between the surface and the solution i.e., at outer Helmholtz plane is called zeta potential [2].

If the membrane polymeric material contains ionisable groups, its nature is reflected in the variation of zeta potential with solution pH. If the groups are weakly acidic then their dissociation takes place gradually and zeta potentials of membrane are expected to be more negative with increasing pH. If the groups are strongly acidic then dissociation takes place very quickly and higher values of negative zeta potential can be expected even at very low pH. On the other hand, presence of basic groups gives positive zeta potential of membrane. Sometimes, membrane materials with no ionic groups also show positive zeta potential at low pH and negative zeta potential at higher pH due to adsorption of H^+ or Na^+ ions and OH^-or Cl^-ions from the solution respectively.

Calculation of Zeta Potential and Surface Charge Density of Membrane

The importance of zeta potential is increasingly recognized as important to the performance of membrane separation as it is responsible for solute-solute or solute-membrane interaction. The zeta potential of a membrane can be determined accurately by measuring streaming potential which is very sensitive to change of solution properties e.g., pH and electrolyte concentration.

When an electrolyte solution is forced to flow through the membrane due to pressure gradient, ions of the diffused double layer are stripped off along the plane of shear. Since the convective flux of the counter ions are greater than that of co-ions, a potential, which is developed across the membrane is called steaming potential. This induced potential causes a back flow of ions until a steady state is reached and the system satisfies the conditions of no current flow. The streaming potential coefficient (V_P) is the ratio of potential difference across the membrane and applied pressure, when the net current is zero.

$$V_P = \left(\frac{\Delta E}{\Delta P}\right)_{i=0} \tag{3.34}$$

Knowing streaming potential coefficient, membrane zeta potential can be calculated using Helmholtz-Smoluchowski Equation [2].

$$\zeta = \frac{V_P \, \mu \, \lambda}{\varepsilon_0 \, D_i} \tag{3.35}$$

Where ζ is the zeta potential, D_i is the dielectric constant of the medium; ε_0 is the permittivity of vacuum, μ and λ the viscosity and conductivity of the bulk solution, respectively. Zeta potential can be calculated using Eq. (3.35) if the following assumptions are satisfied. (i) Potential distribution is independent of pore surface conductivity of liquid in the membrane capillary; (ii) Debye Huckel length, k^{-1} is very small compared to pore radius.

Surface charge density (σ) of membrane can be calculated from the knowledge of zeta potential using the following equation [29].

$$\sigma = 11.74\sqrt{C_1}\ \mathrm{Sinh}\,(19.46\, z\, \zeta) \tag{3.36}$$

and Debye length, which is the double layer thickness, can be calculated using the following equation

$$k^{-1} = 4.31 \times 10^{-10}\left(2C_1\right)^{-0.5} \tag{3.37}$$

3.1.3. Detailed Experiment

3.1.3.1. Ultrafiltration of BSA from Aqueous Solution

Membrane and Materials

BSA (bovine serum albumin, MW 66500) is procured from M/s, SRL Co. (India). A.R grade sodium chloride, sodium hydroxide and hydrochloric acid are obtained from M/s, Merck Limited (India).

Table 3.1. Operating conditions used in experiment

Variable	Operating conditions
BSA concentration in feed (kg/m^3)	0.1, 0.5, 1.0, 1.5
Transmembrane pressure (kPa)	220, 360, 500, 635
Cross flow velocity (m/s)	0.09, 0.12, 0.15, 0.18
Electric field (V/m)	0, 400, 600, 800, 1000
Solution pH	7.4
Electrolyte concentration	1.0 mM

An ultrafiltration membrane (Polyphenylene ethersulfone) of molecular weight cut-off (MWCO) 30 kDa, obtained from M/s, Permionics Membranes Pvt. Ltd., Boroda, Gujrat, (India), has been used for the experiment.

Electro-ultrafiltration Cell and Operating Conditions

Details of electro-ultrafiltration cell have been discussed in section 2.3.1. Electro-ultrafiltration experiments are conducted by taking into account the effect of the four major conditions, namely, pressure, cross-flow velocity, electric field and feed concentration. The details of the operating conditions are shown in Table 3.1

Steps Used during Experiment

Preparation of Feed Solution

Protein solutions are prepared by adding a required amount of powdered bovine serum albumin to an electrolyte solution of ionic strength of 1.0 mM. Electrolyte solutions are prepared by dissolving A.R grade NaCl in doubled distilled water. The pH of the solution is adjusted by drop wise addition of 10 mM NaOH and HCl.

Conduction of Experiments

Details of conduction of experiments are described in section 2.3.2.

Analysis of the Feed and Permeate

The concentrations of BSA in the feed as well as in the permeate and retentate are measured using a Genesys2 Spectrophotometer. For BSA, a wavelength of 280 nm is used and distilled water is taken as a blank. Zeta potential of the feed solutions is measured by a Zetasizer (Malvern Instruments, UK). In the instrument, zeta potential is obtained by measuring electrophoretic mobility and then by using Henry equation. The electrophoretic mobility of the particle is measured by Laser Doppler Velocimetry in the model Nano ZS, ZEN3600. The concentration of proteins used in the measurement of zeta potential, which requires dilute solutions, is 5 kg/m^3 [30].The iso-electric pH of BSA is found to be 4.7 [31]. Zeta potential of BSA increases with increasing pH at a fixed value of ionic strength and increasing ionic strength at constant pH decreases its zeta potential. Zeta potential of BSA at a pH of 7.4 and at an ionic strength of 1mM is measured and found to be -28 mV. Viscosity and conductivity of the samples are measured by Ostwald viscometer and autoranging conductivity meter (Toshniwal Instrument, India), respectively. The variation of pH and conductivity of feed, retentate, permeate solution with applied electric field are reported in Table 3.4. From this table, it can be observed that pH of the permeate increases with electric field. The change of solution pH occurs due to dissociation of water at the electrode. The produced OH^- ions at the cathode are washed out with the permeate, causing an increase of pH of the permeate. Similarly H^+ produced at the anode is responsible for decrease of pH in the retentate. Due to large feed volume and high cross flow velocity, the pH variation of retentate is not significant. At the end of experiment the pH value of retentate decreases from 7.4 to 7.1 causes the zeta potential to increase from -28 mV to -27 mV. Since this change is very small, calculation of electrophoretic mobility is based on zeta potential of initial feed solution.

3.1.3.2. Evaluation of Membrane Surface Charge

Membrane

Two commercially available flat asymmetric ultrafiltration membranes, polyphenylene ethersulfone HFUF-30 (30 kDa) and polyethersulfone PES-20 (20 kDa) obtained from M/s, Permionics Ltd. India are used with pure water permeability of 8.8×10^{-11} m/Pa s and 6.5×10^{-11} m/Pa s respectively.

Table 3.2. Variation of pH and conductivity of feed, retentate and permeate solution during electro-ultrafiltration of BSA solution at pH 7.4, electrolyte strength of 1.0 mM, at a pressure of 360 kPa and a cross flow velocity of 0.12 m/s

E(V/m)	I(mA)	Feed		Retentate		Permeate	
		pH	Conductivity $\times10^3$ (S/m)	pH	Conductivity $\times10^3$ (S/m)	pH	Conductivity $\times10^3$ (S/m)
0	0	7.4	17.4	7.4	17.5	7.4	18.4
400	5	7.4	17.4	7.2	16.9	8.6	19.4
600	10	7.4	17.4	7.2	16.9	9.0	21.0
800	25	7.4	17.4	7.1	16.8	9.5	23.0
1000	40	7.4	17.4	7.1	16.8	9.8	26.0

Preparation of Feed Solution

Electrolyte solutions are prepared by dissolving required quantity of A.R grade NaCl in doubled distilled water (conductivity less than 18 μS /cm). The pH of the BSA in neutral solution is 5.8. The solutions pH is adjusted by drop wise addition of HC1 and NaOH solutions. BSA solutions are prepared by dissolving required quantity of BSA into 1 mM NaCl solution.

Streaming Potential Measurement

Streaming potentials are measured with the following operating conditions: Transmembrane pressure (ΔP): 0 to 2 bar; solution pH: 3.5 to 10; NaCl concentration (C_1): 0.5 mM to 10 mM; cross flow velocity: 0.12 m/s. Before start of experiment (the setup is described in section 2.3), the membranes are soaked for 3 hours in an electrolyte solution having the same ionic strength as used for streaming potential measurement. The electrolyte solution tangentially flows over the membrane surface through a thin channel of 37 cm in length, 3.6 cm in width and 6.5 mm in height. A rotameter in the retentate line measures the flow rate. Pressure inside the flow channel is maintained by operating the bypass valve and measured by a pressure gauge. Permeate is collected from the bottom of the cell. A platinum coated titanium sheet (length 33.5 cm, width 3.4 cm, thickness 1.0 mm) obtained from Ti Anode Fabricators, Chennai (India), mounted in parallel position just above the flow channel acts as an electrode. The membrane is placed over a stainless steel support, which also acts as the other electrode. The electrodes are connected through a high impedance digital micro voltmeter. Due to pressure gradient across the membrane, electrolyte solution flows through the membrane. The effective filtration area is 133.2 cm^2. Initially, streaming potential measurements are carried out at normal pH (i.e., 6.0± 0.2) for each electrolyte concentration. Then pH of the solutions is changed in a stepwise manner. Streaming potential values across the membrane are measured at varying pressures from 0 to 2 bar. The results are confirmed by repeating each experiment at least three times. All experiments are conducted at 30± 2°C.

Zeta Potential and Surface Charge Density of Membrane

Zeta potential (Eq. 3.35), Surface charge density (Eq. 3.36) and Debye length (Eq. 3.37) of HFUF 30 and PES 20 membrane are calculated at different pH and electrolyte concentration and are reported in Tables 3.5 and 3.6. The results show that a decrease in zeta potential with increasing electrolyte concentration. With increase in electrolyte concentration, conductivity of the solution in the membrane pore increases and streaming potential decreases according to Helmohtlz-Smoluchowski equation (Eq. 3.37). Moreover, with increasing electrolyte concentration, Debye length decreases because membrane charge is screened by more counter ions. Therefore, at higher electrolyte concentration it can be assumed that membrane pores behave like an electro neutral channel. It is also observed that for HFUF-30 and PES-20 membranes, the values of zeta potentials are negative for all operating pH (3.5 -10) and gradually become more negative with increasing pH. The negative surface charge is due to adsorption of OH^- or Cl^- ions on the membrane surface from the solution.

Table 3.3. Membrane: PES 20 (20 kDa)

NaCl	pH	Negative zeta potential x10^3 (V)	Negative surface charge ($\mu C/m^2$)	Debye length x 10^9 (m)
1×10^{-3} M	3.5	2.7	120	9.66
	6.0	4.0	290	
	7.0	5.2	370	
	9.0	9.0	650	
	10.0	14.1	1020	

Table 3.4. Membrane: HFUF 30 (30 kDa)

NaCl	pH	Negative zeta potential x 10^3 (V)	Negative surface charge ($\mu C/m^2$)	Debye length x 10^9 (m)
5×10^{-4} M	3.5	2.2	112	13.63
	7.0	6.8	348	
	9.0	8.8	450	
1×10^{-3} M	3.5	1.9	135	9.66
	7.0	4.3	310	
	9.0	7.5	540	
1×10^{-2} M	3.5	1.3	297	3.05
	7.0	3.6	820	
	3.5	4.5	1000	

3.1.4. Osmotic Pressure of Protein Solution

The osmotic pressure of protein solution is assumed to be composed of electrostatic interaction, van der Waals interaction and entropic pressure [18]. Each of these components has been calculated from the formula derived by Bowen et al. [4,18,19], as presented in the appendix. Figure 3.5 (a, b and c) and Fig 3.6 (a, b and c) show the variation of various contributions to the total osmotic pressure with concentration of BSA and Lysozyme at the ionic strength of 1 mM and at different pH. It is evident from the figure that contribution of entropic pressure is more compared to electrostatic interaction, van der Waals interaction to the calculation of total osmotic pressure. Contribution of electrostatic and van der Waals interactions are significant at higher concentration. The total osmotic pressure at the ionic strength of 1mM and for different solution pH is related to protein concentration by a polynomial fit and the results are presented in the appendix.

3.1.4.1. Effect of pH and Solute Concentration on Osmotic Pressure of BSA

Figure 3.4 (a, b and c) show the variation of various contributions to the total osmotic pressure with BSA concentration for different pH and at the ionic strength of 1mM.

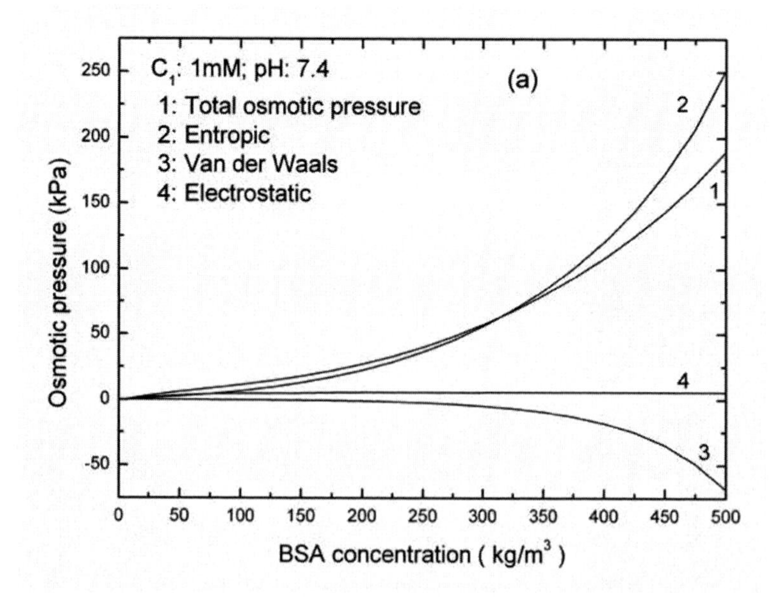

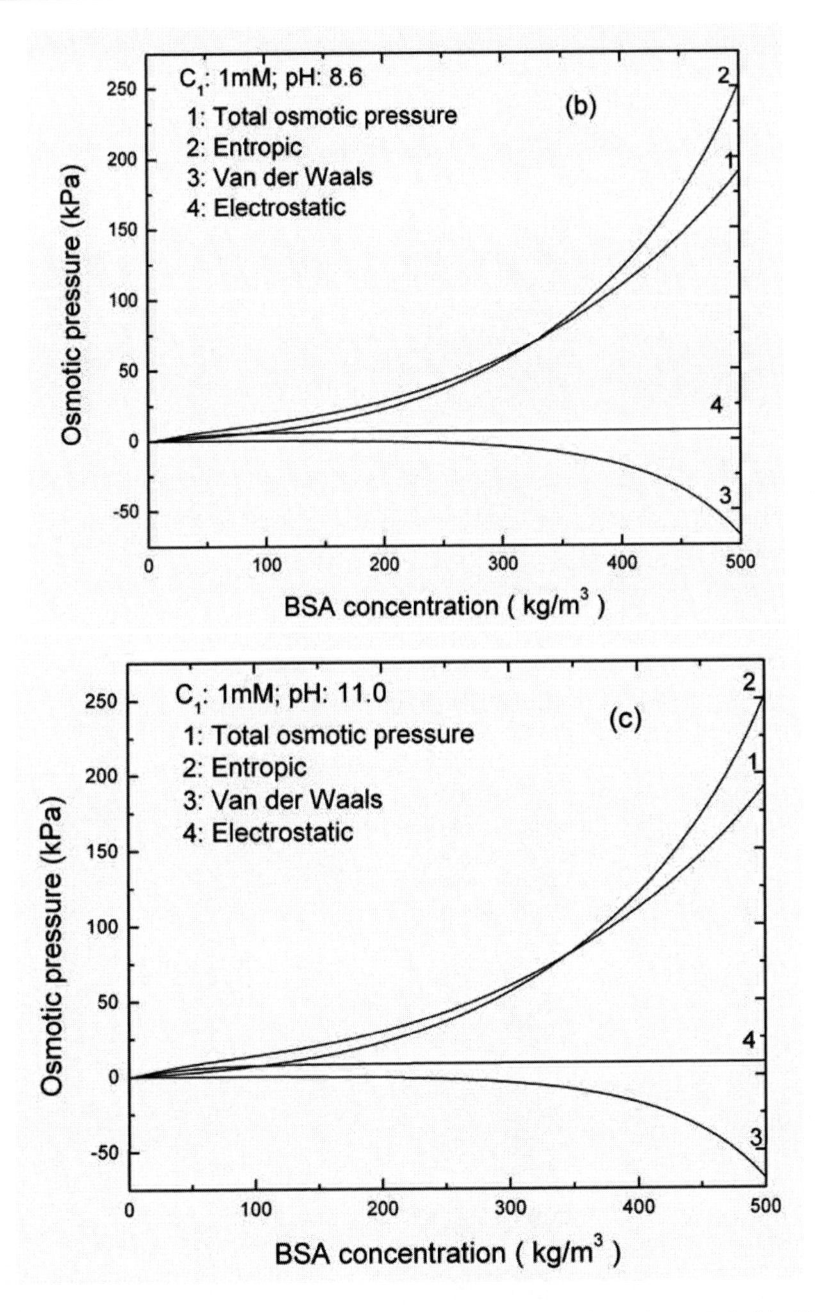

Figure 3.4. Variations of osmotic pressure with BSA concentration (a) pH: 7.4, (b) pH: 8.6, (c) pH: 11.0.

3.1.4.2. Effect of pH and Solute Concentration on Osmotic Pressure of Lysozyme

Figure 3.5 (a, b and c) shows the variation of various contributions to the total osmotic pressure with Lysozyme concentration for different pH and at the ionic strength of 1mM.

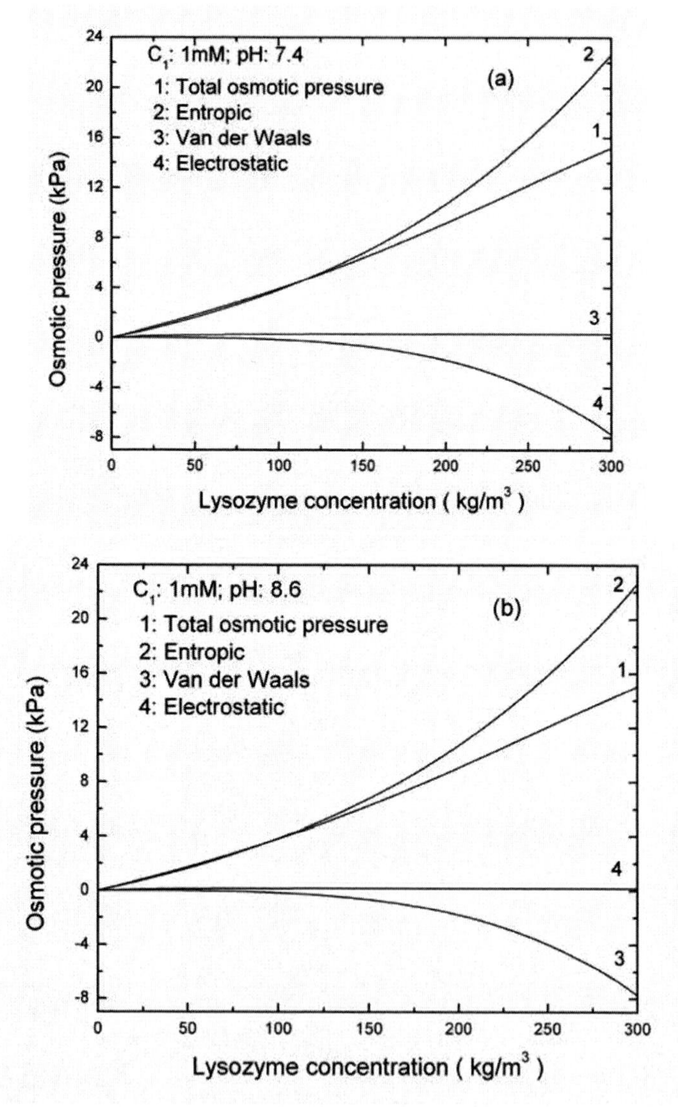

Figure 3.5 (Continued)

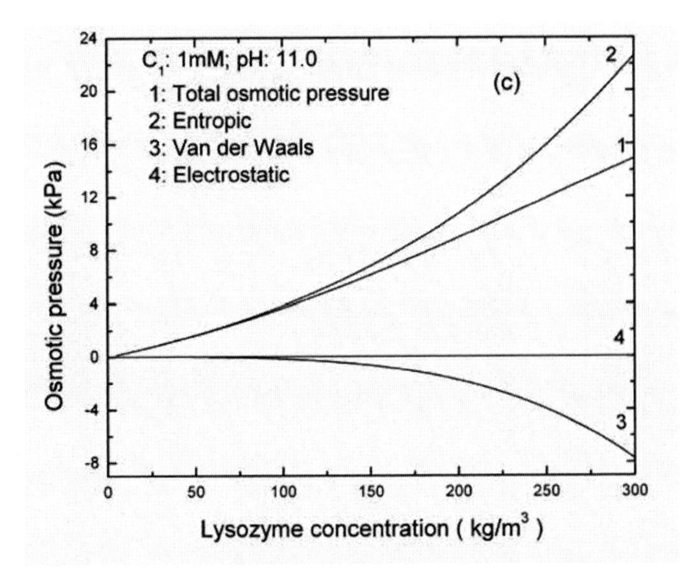

Figure 3.5. Variations of osmotic pressure with Lysozyme concentration (a) pH: 7.4, (b) pH: 8.6, (c) pH: 11.0.

3.1.5. Quantification of Permeate Flux and Membrane Surface Concentration

3.1.5.1. Effect of Electric Field on Concentration Boundary Layer Thickness

Figure 3.6 shows the development of concentration boundary layer thickness (δ_1) along the dimensionless length of the channel (x*) for different electric field. As is observed from the figure that the development of concentration boundary layer is very rapid during the initial part of the channel and gradual thereafter. It may also be observed that, at any position of the flow channel, with increase in electric field, concentration boundary layer thickness increases. For example, at the exit of the channel, by applying electric field 1000 V/m, δ_1 increase from 0.098 mm to 0.142 mm (i.e., 8.7% of the channel height) compared to zero electric field while other operating conditions are held constant.

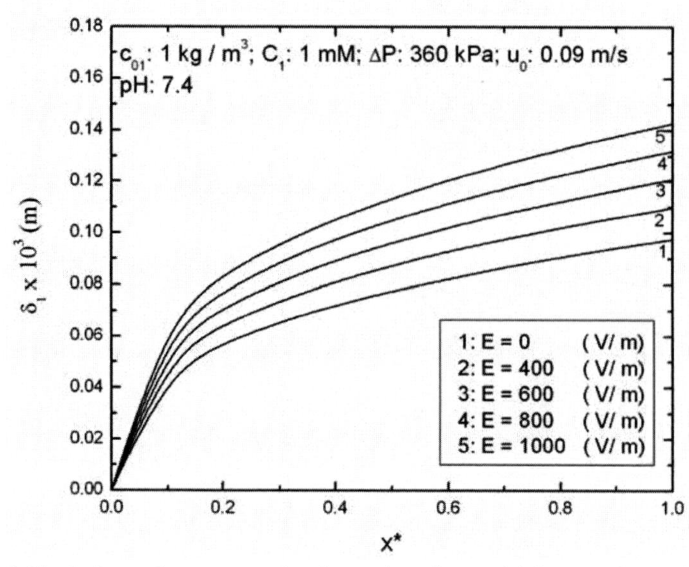

Figure 3.6. Variation of concentration boundary layer thickness along the length of the channel.

3.1.5.2. Effect of Electric Field on the Variation of Membrane Surface Concentration and Permeate Flux along the Length of the Channel

The variations of membrane surface concentration (c_{m1}) and permeate flux (v_w) along the length of the channel for different electric fields are illustrated in Figures 3.7 and 3.8. It can be observed from the figure that for a fixed value of electric field, permeate flux decreases very sharply at the entrance of the channel followed by a smoother and slower decline towards the exit of the channel. This is a direct consequence of the rapid growing membrane surface concentration (Figure 3.7) and concentration boundary layer (Figure 3.6) close to the entrance and gradual growth for the rest of the channel. For example, at the end of the channel, for a fixed value of feed concentration of 1 kg/m^3, cross flow velocity of 0.09 m/s and pressure of 360 kPa, c_{m1} decreases from 555 kg/m^3 to 335 kg/m^3 (Figure 3.7) and permeate flux increases from 27.2 L/ m^2 h to 93.1 L/ m^2 h (Figure 3.8) when an electric field strength of 1000 V/m is applied. It is also evident from the figure that at any particular channel position, higher values of permeate flux is obtained for higher electric field. Under application of d.c. electric field, the negatively charged protein molecules move toward the anode away from the membrane surface. Therefore membrane surface concentration decreases resulting in an increase in permeate flux.

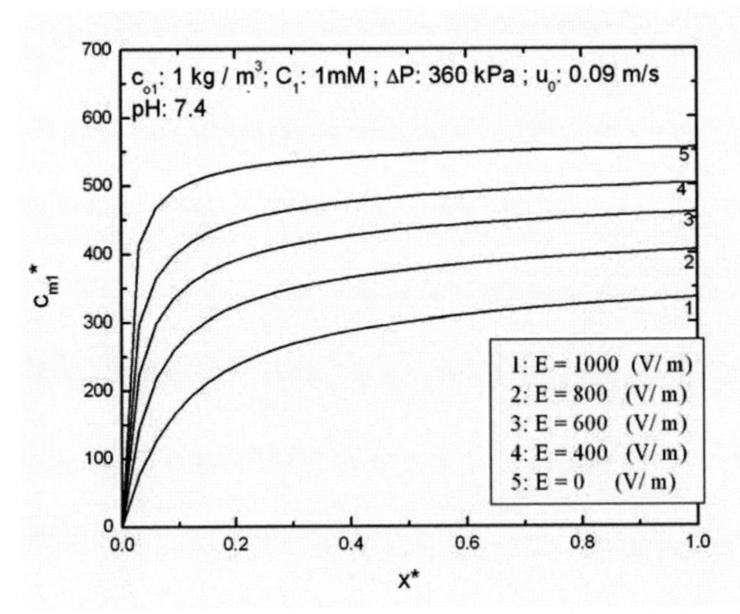

Figure 3.7. Variations of membrane surface concentration along the length of the channel for different electric field.

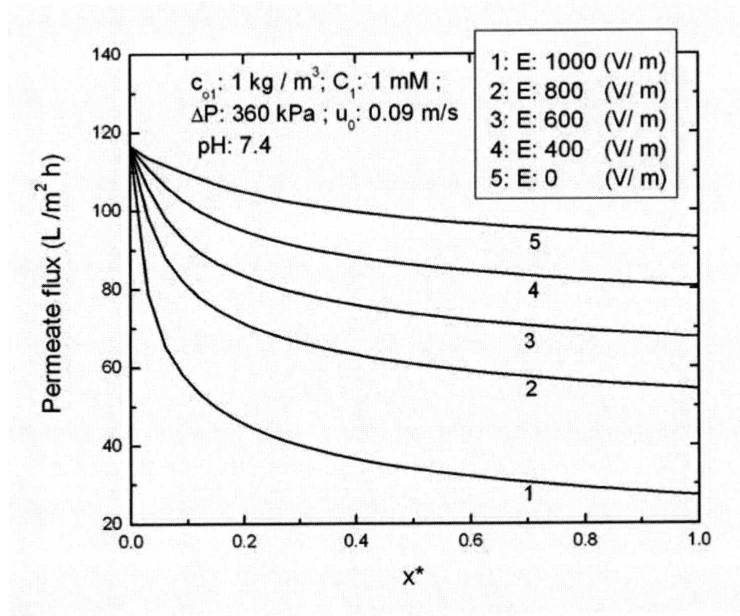

Figure 3.8. Variations of permeate flux profile along the length of the channel for different electric field during ultrafiltration of BSA from aqueous solution.

3.1.5.3. Effect of Electric Field on the Membrane Surface Concentration and Permeate Flux

Figures 3.9 and 3.10 are the plots of average values of permeate flux and membrane surface concentration against electric field, respectively, using feed concentration as a parameter. For a fixed feed concentration, permeate flux increases with increase in electric field as expected. Under application of d.c. electric field, electrophoretic movement of proteins toward the anode resists them from their accumulation on the membrane surface.

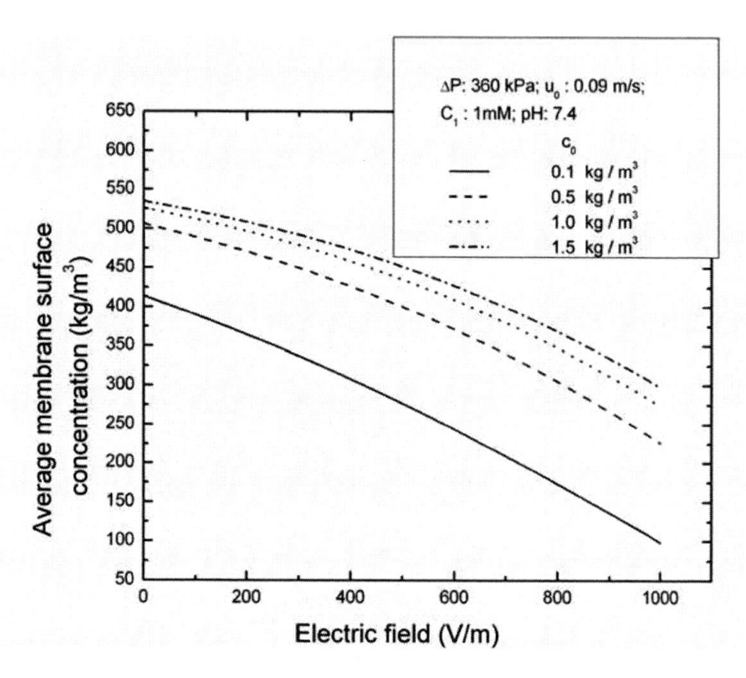

Figure 3.9. Variations of average membrane concentration with electric field for different feed concentration during ultrafiltration of BSA from aqueous solution.

Therefore, with increase in electric field membrane surface concentration decreases (Figure 3.9) which in turn decreases the osmotic pressure and hence net driving force for solvent flow increases resulting in an increase in permeate flux (Figure 3.10). Figure 3.10 also demonstrates the comparison between the experimental data and model prediction under the same operating conditions. As can be observed from the figure, most of the experimental data lie within ±10% of the model predicted values of average permeate flux. It can be seen from the figure that for a fixed feed BSA concentration of 1.0 kg/m^3, at 1000 V/m, average permeate flux is about 99.7 L/ m^2 h whereas it is about 39.25 L/ m^2 h for without electric field. It can also observed from this figure that for a fixed value of electric field strength the permeate flux decreases with increase in feed concentration. For a particular electric field, with increase in feed concentration, membrane surface concentration increases, resulting in higher osmotic pressure and hence causes a reduction in the permeate flux.

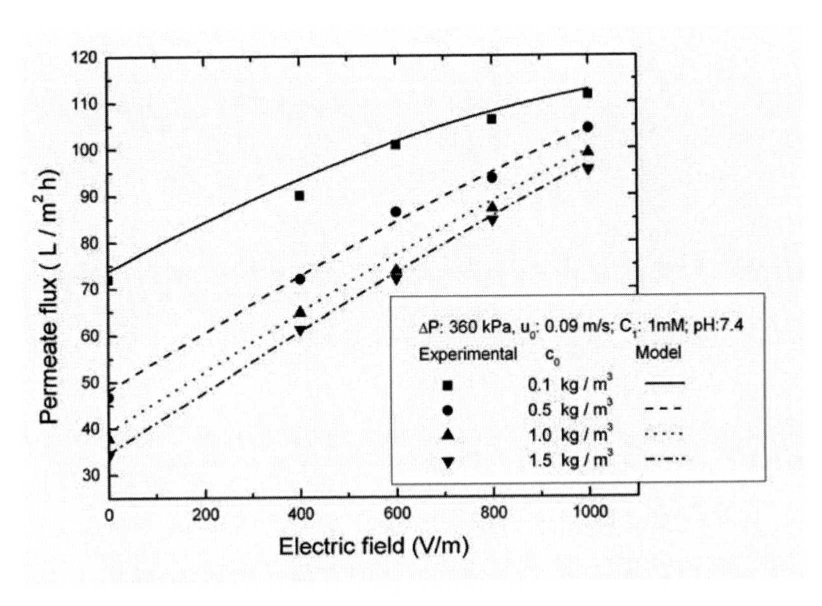

Figure 3.10. Variations of average permeate flux with electric field for different feed concentration during ultrafiltration of BSA from aqueous solution.

3.1.5.4. Effect of Pressure on the Permeate Flux

Fig 3.11 shows the variation of permeate flux with transmembrane pressure for different values of feed concentration. It is observed from the figure that flux increases almost linearly for lower pressure range but at higher pressure the increment is gradual. At higher pressure, concentration polarization becomes severe and membrane surface concentration increases which in turn increases osmotic pressure. This reduces the net driving force for solvent flow resulting in a gradual increase of permeate flux. For a fixed pressure, permeate flux is lower for higher feed concentration. This can be explained by the fact that membrane surface concentration increases due to severe concentration polarization. This increases the osmotic pressure resulting in a reduction of driving force for permeate flux. Hence, permeate flux decreases with feed concentration.

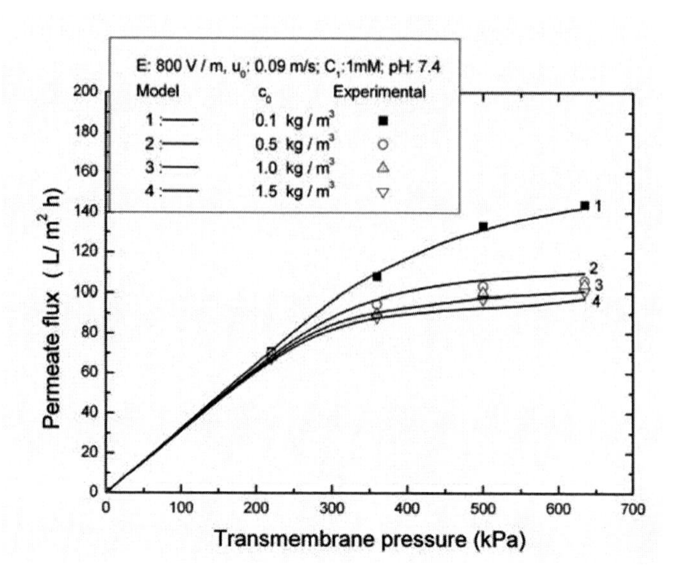

Figure 3.11. Variations of average permeate flux with transmembrane pressure for different feed concentrations during ultrafiltration of BSA from aqueous solution.

3.1.5.5. Effect of Cross Flow Velocity on Permeate Flux

The effect of cross flow velocity on the permeate flux are presented in Figures 3.12 (a and b). According to these figures, permeate flux increases with increase in cross flow velocity. Similar trends are observed with and without electric field cases. This can be explained by the fact that c_{m1} is lowered at higher cross flow velocity due to forced convection imposed by cross flow velocity, leading to an increase in concentration gradient between the membrane surface and the bulk and a lowering of c_{m1}. Therefore, permeate flux increases with increase in cross flow velocity. It can further be noticed that effect of cross flow velocity is more pronounced at zero electric field compared to zero electric field due to less membrane concentration in presence of electric field. For example, at a feed concentration of 1.0 kg/m^3, increase in cross flow velocity from 0.06 m/s to 0.18 m/s causes to increase in permeate flux from 34.7 L/ m^2 h to 47.9 L/ m^2 h at 0 V/m (i.e., 38% increase) while at 1000 V/m, permeate flux increases from 97.2 L/ m^2 h to 104.4 L/ m^2 h (i.e., 7.4% increase) keeping other operating conditions unaltered.

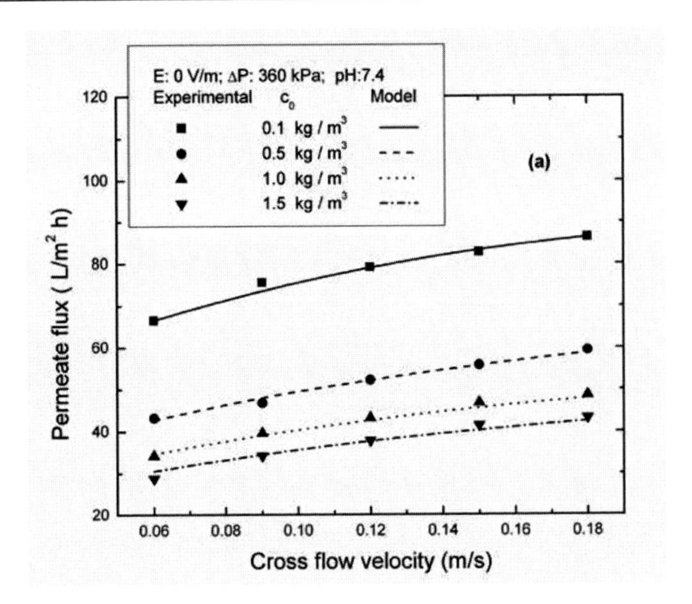

Figure 3.12. (Continued)

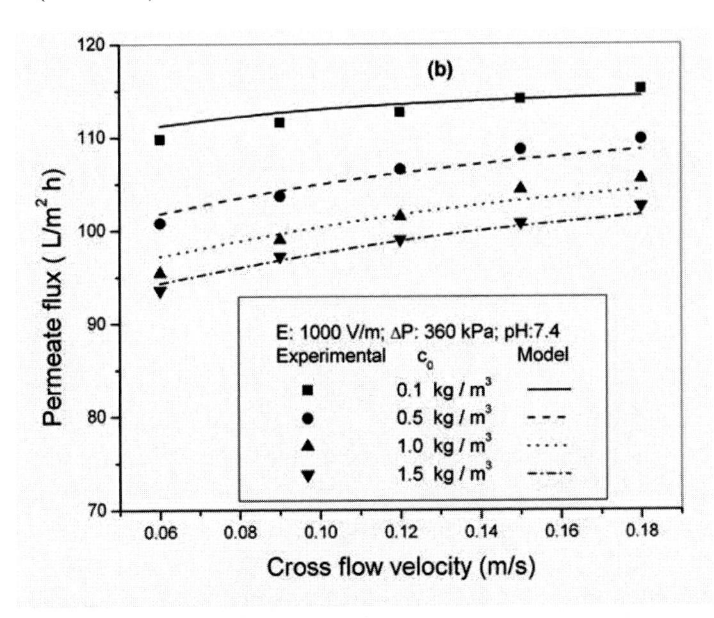

Figure 3.12. Variations of average permeate flux with cross flow velocity for different feed concentration during ultrafiltration of BSA from aqueous solution (a) at 0 V/m, (b) at 1000 V/m.

3.1.5.6. Variation of Mass Transfer Coefficient

The variations of Sherwood number along the length of the flow channel at different electric field are described in Figure 3.13. It is observed from the figure that local Sherwood number decreases downstream of the flow channel.

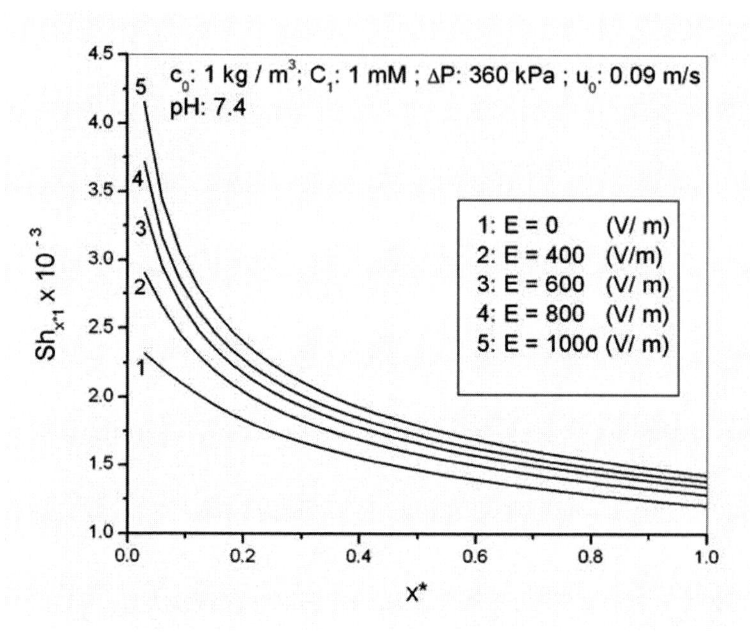

Figure 3.13. Variations of Sherwood number along the length of the channel for different electric field during ultrafiltration of BSA from aqueous solution.

This decrease is very sharp near the entrance of the channel and gradual in the later part. The mass transfer coefficient varies with the thickness of concentration boundary layer as $k_1 = D_1/\delta_{c1}$. As observed from Figure 3.6, the concentration boundary layer develops very sharply close to the channel entrance and the development is gradual later on. Consequently, the mass transfer coefficient decreases rapidly at the channel entrance and the decrease become gradual at the down stream of the channel. From this figure, it is also clear that with increase in applied electric field local Sherwood number decreases. In presence of electric field, concentration boundary layer thickness increases (Figure 3.6) due to electrophoresis. Hence, mass transfer coefficient and consequently Sherwood number decreases with an increase in electric field. For example, at the exit of the channel, for feed concentration of 1 kg/m^3, cross flow velocity of 0.09 m/s and a pressure of 360 kPa, Sherwood number decreases from 1432 to 1208 when electric field of 1000 V/m is applied.

3.1.5.7. Effect of Solution pH during Ultrafiltration of BSA from Aqueous Solution

The effects of solution pH on permeate flux during ultrafiltration of BSA solution through fully retentive membrane are shown in Figure 3.14.

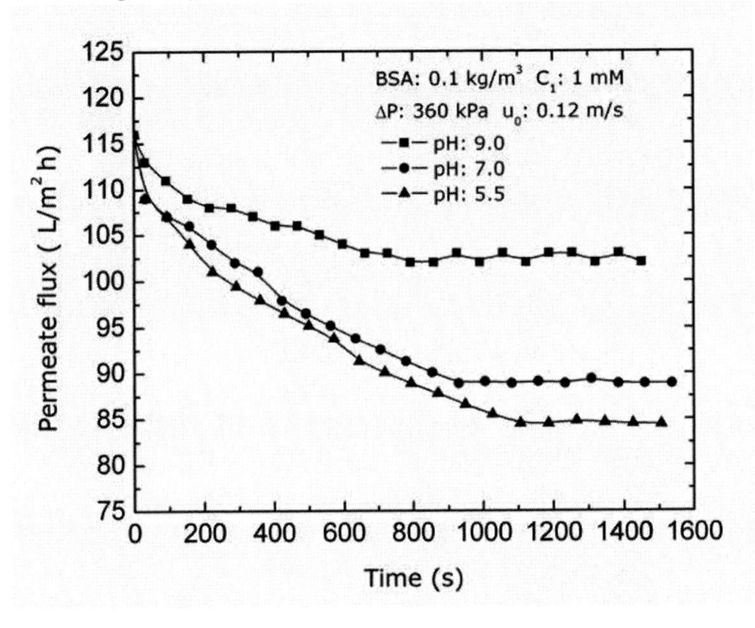

Figure 3.14. Variation of permeate flux with time at different solution pH.

It is clear from the figure that, permeate flux decreases with time of operation. During experiments the membrane surface concentration keeps on increasing leading to an increase in osmotic pressure near the membrane-solution interface. This reduces the available driving force for solvent flow. Therefore, permeate flux decreases gradually with time. Experimental results confirm that, at pH 9.0, during operation permeate flux declines about 12% from its initial value, whereas, it is about 27% at pH 5.5.

It is also observed from the figure that with increase in pH from 5.5 to 9.0 steady state permeate flux increases from 84.4 L/m^2 h to 102 L/m^2 h. Electrostatic interaction between solute-solute and solute-membrane plays an important role during ultrafiltration of charged particles. Zeta potential of the BSA solutions at pH 5.5, 7.0 and 9.0 are measured and found to be -18 mV, -26.5 mV and -33.0

mV respectively. At pH 5.5 (isoelectric point of BSA is 4.7), electrostatic interaction between BSA and membrane surface is very small resulting in maximum accumulation of protein near the membrane surface. Therefore, at pH 5.5, the concentration polarization is more which in turn decreases the net driving force for solvent flow and hence permeate flux decreases. With increase in solution pH, zeta potential of protein as well as membrane increases. At pH 7 and 9.0, both BSA and membrane become more negative. Strong electrostatic repulsion between both solute-solute and solute membrane restricts the concentration polarization over the membrane surface and hence permeate flux increases. At alkaline pH, although the electrostatic repulsion between membrane surface and BSA increases, it is not sufficient enough to completely overcome concentration polarization. For example, at pH 9.0, the steady state flux is about 12% less than the initial flux.

3.2. Electric Field Enhanced Fractionation of BSA and Lysozyme

The protein-protein and protein-membrane interactions largely depend on physico-chemical properties of solution such as pH and ionic strength of the solution. Hence, the transmission of a solute through the membrane is a strong function of pH, ionic strength and operating conditions such as pressure, cross flow velocity etc. The importance of these parameters on the permeate flux and the selectivity of protein fractionation has been described by several investigators. Therefore, to increase the flux as well as selectivity of solute separation, the proper control of the solution pH, the ionic strength, and the membrane protein electrostatic interaction and concentration polarization is necessary. It is normally observed that highest protein transmission occurs at its iso-electric point. van Eijndhoven et al. [32] have improved the selectivity of albumin and hemoglobin by varying salt concentration and adjusting pH to near iso-electric point of hemoglobin. Millesime et al. [33] have investigated the effect of ionic strength on fractionation of BSA and Lysozyme using inorganic membrane. Iritani et al. [34] have discussed the influence of pH and ionic strength on the transmission of Lysozyme during fractionation of BSA and Lysozyme using polysulphone membrane. Saksena and Zydney [35] demonstrated the importance of electrostatic interactions on protein filtration during separation of bovine serum albumin (BSA) from immunoglobulin (IgG) using an Amicon stirred ultrafiltration cell. Ehsani et al. [36] have shown the effect of pH, salt concentration and membrane surface modification on permeate flux, total protein retention and membrane zeta potential in ultrafiltration of chicken egg white proteins. The effect of salt concentration and BSA-Lysozyme interaction on the permeation of Lysozyme is discussed by Ingham et al. [37]. Besides, liquid pulsation [38], pressure pulsation [39], ultrasound have a significant effect on both flux and selectivity of protein fractionation. Teng et al. [40] have investigated the effect of ultrasound on separation ability, flux and protein structure during cross flow ultrafiltration of binary protein solution of BSA and Lysozyme.

In this section the application of electric field is explored during the fractionation of binary protein solution of BSA and Lysozyme. During fractionation of two proteins with different pH, suitable choice of operating pH can make the protein oppositely charged. The polarity of electric field is selected in such a way so that the larger protein molecules move in a direction away from the membrane surface and consequently facilitates the transmission of smaller protein molecules, resulting in higher flux and separation. The effects of external d.c. electric field, solution pH, operating parameters, e.g., pressure difference, feed composition and cross-flow velocity, on permeate flux and Lysozyme retention during fractionation of protein mixture of BSA (M.W: 69000 and iso-electric point 4.7) and Lysozyme (M.W: 14600 and iso-electric point 11.0) are studied by ultrafiltration using a 30 kDa membrane.

3.2.1. Theoretical Aspects

Electric field assisted fractionation of binary protein solution of BSA and Lysozyme is assumed to be osmotic pressure governed.The flow geometry of the electric field enhanced ultrafiltration is presented in Figure 3.15. The following assumptions are made for the model development: (i) laminar, steady state flow, (ii) velocity boundary layer is completely developed (small entrance length); (iii) permeate velocity is small compared to the axial cross flow velocity (iv) no adsorption of solute over the membrane surface; (v) the channel height is sufficiently small so that electric field is assumed to be uniform; (vi) the flow is two dimensional inside the boundary layer. (vii) physical properties of the components are assumed to constant.

During ultrafiltration, solutes are convected by the pressure gradient towards the membrane surface. The rejected solutes tend to accumulate over the membrane surface. Since the membrane surface concentration is higher than the bulk concentration, back diffusion of rejected solutes from the surface towards the bulk of the solution takes place. In presence of an external d.c. electric field (with the top surface being the positive electrode), the particles move away from the membrane surface due to electrophoresis. At steady state, a material balance for the BSA (component1) in the concentration boundary layer results the following equation:

$$D_1 \frac{dc_1}{dy} + (v_w - v_{e1})\, c_1 = 0 \tag{3.38}$$

The pertinent boundary conditions are,

at $y = 0$, $c = c_{m1}$ (3.39)

at $y = \delta_{c1}$, $c = c_{01}$ (3.40)

where, δ_{c1} is the thickness of the concentration boundary layer for BSA and v_{e1} is the electrophoretic velocity which is related with electrophoretic mobility as,

$$v_{e1} = u_{e1} E \tag{3.41}$$

The electrophoretic mobility, u_{e1} can be expressed using Helmholtz-Smoulochowski's equation [2] as,

layer has been greatly enlarged for clarity.

$$u_{e1} = \frac{\varepsilon_0 D_i \xi_{eff}}{\mu} \tag{3.42}$$

Eq. (3.38) can be solved along with the boundary conditions, i.e. Eqs. (3.39) and (3.40) and can be expressed as:

$$v_w = k_1 \ln \frac{c_{m1}}{c_{01}} + v_{e1} \tag{3.43}$$

where k_1 ($= D_1/\delta_{c1}$) is the mass transfer coefficient of BSA molecule.

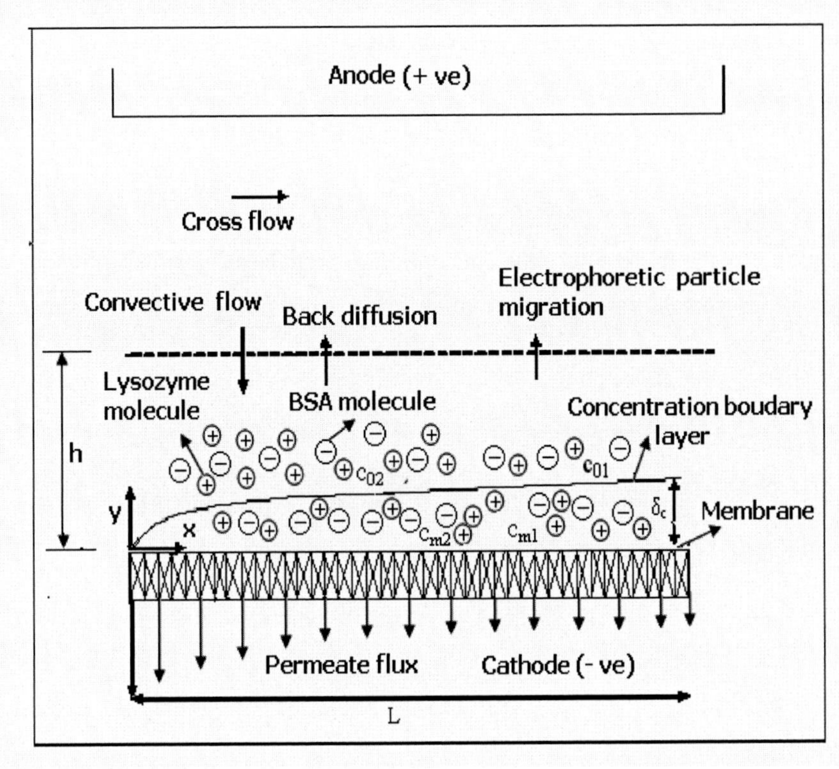

Figure 3.15. Schematic of formation of external concentration boundary layer over the membrane surface in presence of external d.c. electric field. The thickness of the boundary

The values of zeta potential at various pH conditions for pure components and the mixture of BSA and Lysozyme are presented in Table 2. It is observed from this table that the zeta potential of BSA is always negative for the pH values studied herein, whereas, those of Lysozyme are positive (or zero). From the last column of Table 2, it is observed that BSA-Lysozyme mixture is always negative with absolute values less that for pure BSA. The material balance for free Lysozyme (Lysozyme not in complex form with BSA) in the concentration boundary layer results,

$$D_2 \frac{dc_2}{dy} + v_w (c_2 - c_{p2}) = 0 \tag{3.44}$$

The pertinent boundary conditions are,

$$\text{at } y = 0 \quad c = c_{m2} \tag{3.45}$$

$$\text{at } y = \delta_{c2}, \quad c = c_{02} \tag{3.46}$$

where, δ_{c2} is the thickness of the concentration boundary layer for Lysozyme and v_{e2} is the electrophoretic velocity which is related with electrophoretic mobility as,

$$v_{e2} = u_{e2} E \tag{3.47}$$

The electrophoretic mobility, u_{e2} can be expressed using Helmholtz-Smoulochowski's equation [2] as,

$$u_{e2} = \frac{\varepsilon_0 D_i \xi_2}{\mu} \tag{3.48}$$

Eq. (3.44) can be solved along with the boundary conditions, i.e., Eqs. (3.45) and (3.46) and can be expressed as:

$$v_w = k_2 \ln \frac{c_{m2} - c_{p2}}{c_{02} - c_{p2}} - v_{e2} \tag{3.49}$$

where, k_2 (= D_2/δ_{c2}) is the mass transfer coefficient of Lysozyme molecule.

Leveque relationship for laminar flow regime without suction is used for estimation of mass transfer coefficient.

$$\text{where, } k_1 = \frac{1.86\, D_1}{d_e}\left(\text{Re}\, \text{Sc}_1 \frac{d_e}{L}\right)^{\frac{1}{3}} = 1.86\left(\frac{u_o\, D_1^2}{d_e L}\right)^{\frac{1}{3}} \tag{3.50}$$

$$\text{and } k_2 = \frac{1.86\, D_2}{d_e}\left(\text{Re}\, \text{Sc}_2 \frac{d_e}{L}\right)^{\frac{1}{3}} = 1.86\left(\frac{u_o\, D_2^2}{d_e L}\right)^{\frac{1}{3}} \tag{3.51}$$

where, d_e is the equivalent diameter of the flow channel. For a thin channel, the value of d_e is 4h, h is the half height of the channel.

In an osmotic pressure governed ultrafiltration, solvent flow through the porous membrane can be expressed by the phenomenological equation as,

$$v_w = L_p\ (\Delta P - \Delta\pi) \tag{3.52}$$

where, $\Delta\pi = \pi_m - \pi_p$

For two component system, $\Delta\pi$ is expressed as,

$$\Delta\pi = \Delta\pi_1 + \Delta\pi_2 + B' c_{m1} c_{m2} \tag{3.53}$$

In the above equation, first two terms on right hand side correspond to the pure components and third term indicates interaction between protein molecules which is a function of solution pH and ionic strength. Osmotic pressure difference for BSA is expressed as,

$$\Delta\pi_1 = \pi_1\big|_{cm1} - \pi_1\big|_{cp1} \tag{3.54}$$

Since BSA is completely retained by the membrane, $c_{p1} = 0$ and so, $\pi_1\big|_{cp1} = 0$

Therefore, osmotic pressure difference across the membrane for BSA becomes

$$\Delta\pi_1 = a_1 c_{m_1} + a_2 c_{m1}^2 + a_3 c_{m1}^3 + a_4 c_{m1}^4 \tag{3.55}$$

Similarly, osmotic pressure difference for Lysozyme is,

$$\Delta\pi_2 = \pi_2\big|_{cm2} - \pi_2\big|_{cp2} \tag{3.56}$$

$$= b_1\left(c_{m2} - c_{p2}\right) + b_2\left(c_{m2}^2 - c_{p2}^2\right) + b_3\left(c_{m2}^3 - c_{p2}^3\right) + b_4\left(c_{m2}^4 - c_{p2}^4\right) \tag{3.57}$$

The methodology for calculating the osmotic pressure of BSA and Lysozyme protein at a particular pH and salt concentration is adopted from Bowen et al. [4,18,19] and is discussed in appendix. Osmotic pressure correlations of BSA solution and Lysozyme solution are given in appendix.

The membrane-solute system is characterized by the parameter real retention (R_r), which is constant for a particular membrane-solute combination. It may be mentioned here that between the two proteins studied herein (BSA and Lysozyme), BSA is completely retained by 30 kDa membrane whereas Lysozyme is partially retained. Hence, R_{r2} of Lysozyme is defined as,

$$R_{r2} = 1 - \frac{c_{p2}}{c_{m2}} \tag{3.58}$$

The observed retention (R_{o2}) relates the permeate concentration with feed bulk concentration as,

$$R_{o2} = 1 - \frac{c_{p2}}{c_{o2}} \tag{3.59}$$

where, c_o, c_p and c_m are the bulk, permeate and membrane surface concentration of the solute.

Using Eqs. (3.57) and (3.58), the expression of $\Delta\pi_2$ becomes

$$\Delta\pi_2 = b_1 c_{m2} R_{r2} + b_2 c_{m2}^2 \left[1-(1-R_{r2})^2\right] + b_3 c_{m2}^3 \left[1-(1-R_{r2})^3\right] + b_4 c_{m2}^4 \left[1-(1-R_{r2})^4\right] \tag{3.60}$$

Thus, Eq. (3.52) is expressed as,

$$v_w = Lp \left(\begin{array}{l} \Delta P - a_1 c_{m_1} - a_2 c_{m1}^2 - a_3 c_{m1}^3 - a_4 c_{m1}^4 - b_1 c_{m2} R_{r2} - \\ b_2 c_{m2}^2 \left[1-(1-R_{r2})^2\right] - b_3 c_{m2}^3 \left[1-(1-R_{r2})^3\right] - \\ b_4 c_{m2}^4 \left[1-(1-R_{r2})^4\right] - B' c_{m1} c_{m2} \end{array} \right) \tag{3.61}$$

Combining Eqs. (3.49) and (3.58) the expression of flux is

$$v_w = k_2 \ln \frac{c_{m2} R_{r2}}{c_{02} - c_{m2}\left(1 - R_{r2}\right)} - v_{e2} \tag{3.62}$$

Combining Eqs. (3.43) and (3.62), a connecting equation between c_{m1} and c_{m2} is obtained,

$$k_1 \ln \frac{c_{m1}}{c_{01}} + v_{e1} = k_2 \ln \frac{c_{m2} R_{r2}}{c_{02} - c_{m2}\left(1 - R_{r2}\right)} - v_{e2} \tag{3.63}$$

From Eq. (3.63), c_{m2} can be expressed in terms of c_{m1}

$$c_{m2} = \frac{c_{o2}}{\left[1 - R_{r2}\left(1 - F\right)\right]} \tag{3.64}$$

where, $F = \dfrac{1}{\left(\dfrac{c_{m1}}{c_{o1}}\right)^{\frac{k1}{k2}} \exp\left(\dfrac{v_{e1} + v_{e2}}{k_2}\right)}$

For a known set of operating conditions, values of permeate flux and permeate concentration can be estimated by simultaneous solution of Eqs. (3.43), (3.49), (3.53) and (3.59), if the values of parameters (B' and R_{r2}) known a priori. In the present work, these parameters values are estimated by an optimization technique using BCPOL subroutine of IMSL math library utilizing direct search algorithm. The algorithm for numerical calculation is presented in Figure 3.16. The results obtained are described in the subsequent sections.

Algorithm for Numerical Calculation

For given input parameters such as system geometry (d_e, L), operating conditions (ΔP, u_0, c_{o1}, c_{o2}, pH), physical properties (D_1, D_2), osmotic pressure coefficients (a_1, a_2, a_3, a_4, b_1, b_2, b_3, b_4)

1. R_{r2} and B' are Guessed
2. c_{m1} is guessed

3. c_{m2} is calculated from Eq. (3.64)
4. v_w is calculated from Eq. (3.61)
5. c_{m1} is calculated from Eq. (3.43)
6. $\left| c_{m1,cal} - c_{m1,gause} \right| \leq \varepsilon$ is checked
7. If yes, c_{p2} is calculated from Eq. (3.58)
8. If no, another value of c_{m1} is guessed and steps from 3-6 are repeated.
9. $\sum_i \left(\frac{v_{w,exp} - v_{w,cal}}{v_{w,exp}}\right)^2 + \left(\frac{c_{p2,exp} - c_{p2,cal}}{c_{p2,exp}}\right)^2 \leq \varepsilon$ is checked
10. If yes, B', R_{r2}, v_w, c_{m1}, c_{m2}, c_{p2} are the values for this operating conditions.
11. If no, another values of R_{r2} and B' guessed and steps from 2-8 are repeated.

3.2.2. Detailed Experiment

3.2.2.1. Membrane and Materials

BSA and Lysozyme (from chicken egg white) were purchased from M/s, SRL Pvt. Ltd., India. Physical properties of the two proteins are shown in Table 3.2. Protein solutions are prepared by adding a required amount of powdered protein to an electrolyte solution of ionic strength of 1.0 mM. Electrolyte solutions are prepared by dissolving A.R grade NaCl (obtained from MERCK India Ltd.) in double distilled water. pH of the solution is adjusted by drop wise addition of 10mM NaOH and HCl. Since the objective of the study is to obtain pure Lysozyme in the permeate stream, the membrane is chosen in such a way so that higher molecular weight protein (BSA) is fully retained by the membrane

Table 3.5. Properties of model proteins used in experiments

Characteristics	Bovine serum albumin (BSA)	Lysozyme (LYS)
Molecular weight (kDa)	69 [41]	14.6 [42]
Dimension (nm^3)	11.6 x 2.7 x 2.7 [43]	4.5 x 3 x 3 [43]
Hydrodynamic radius (nm)	3.13 [41]	2.09 [42]
Specific volume (cc/gm)	0.734 [18]	0.703 [42]
Diffusion coefficient (m^2/s)	6.9×10^{-11} [44]	11.8×10^{-11} [45]
Isoelectric point	4.7 [46]	11.0 [46]
Hamaker constant (J)	0.753×10^{-20} [19]	$7.7\ K_B\ T$ [42]

Hence an ultrafiltration membrane (Polyphenylene ethersulfone) of molecular weight cut-off (MWCO) 30 kDa, obtained from M/s, Permionics Membranes Pvt. Ltd., Boroda, Gujrat, (India), has been selected for the experiment.

Table 3.6. Operating conditions used in experiments

Variable	Operating conditions
Lysozyme to BSA concentration ratio in feed ($kg.m^{-3}$/ $kg.m^{-3}$)	0.1: 0.05; 0.1:0.1; 0.1:0.2; 0.1:0.3
Transmembrane pressure (kPa)	220, 360, 500, 635
Cross flow velocity (m/s)	0.09, 0.12, 0.15, 0.18
Electric field (V/m)	0, 400, 600, 800, 1000
Solution pH	7.4, 8.6, 11.0
Electrolyte concentration	1.0 mM

3.2.2.2. Electro-ultrafiltration Cell

From the feed tank, feed solution is pumped and allowed to flow tangentially over the membrane surface through a thin channel of 37 cm in length, 3.6 cm in width and 4.5 mm in height. The detailed description of the cell is presented in section 2.3.

3.2.2.3. Experimental Design

Electro-ultrafiltration experiments are conducted by taking into account the effect of the four operating variables, namely, pressure, cross-flow velocity, electric field, feed concentration and solution pH. The ranges of the operating conditions are shown in Table 3.6.

3.2.2.4. Procedure

Conduction of Experiments

Details of conduction of experiments are described in section 2.3.

Analysis of the Feed and Permeate

The concentrations of Lysozyme in the permeate are measured using a Genesys2 Spectrophotometer. For BSA and Lysozyme, a wavelength of 280 nm is used and distilled water is taken as blank. During electric field enhanced ultrafiltration of BSA solution, the concentration of BSA in the permeate stream is found to be nil and the permeate stream contains only Lysozyme. Zeta potentials of the feed solutions are measured by a Zetasizer (Malvern Instruments, UK) and are shown in Table 3.3. In the instrument, zeta potential is obtained by measuring electrophoretic mobility and then using Henry equation. The electrophoretic mobility of the particle is measured by Laser Doppler Velocimetry in the model Nano ZS, ZEN3600. Viscosity and conductivity of the samples are measured by Ostwald viscometer and autoranging conductivity meter (Toshniwal Instrument, India), respectively.

Table 3.7. Effect of pH on zeta potential of model proteins in 1 mM NaCl solution

	Zeta potential (mV)		
pH	BSA	LYS	BSA + LYS
7.4	-28	6.7	- 7.1
8.6	-30	5.0	-12.0
11.0	-33	0.0 [46]	-29.0

3.2.3. Quantification of Permeate Flux and Solute Retention

3.2.3.1. Effect of Pressure on Permeate Flux and Observed Retention during Ultrafiltration of Single Protein Solution

Figure 3.16 shows a plot of permeate flux and observed retention as a function of pressure during ultrafiltration of single protein solution of BSA and Lysozyme. During ultrafiltration of aqueous solution of BSA, steady state flux increases linearly for lower pressure and gradually for higher pressure values. At higher pressure, concentration polarization on the membrane surface becomes severe. As a result, osmotic pressure of the solution near the surface increases which reduces the driving force of solvent flow and therefore, the permeate flux increases at a slower rate at higher pressure. For all operating pressures, BSA is found to be fully retentive. During ultrafiltration of Lysozyme, permeate flux increases almost linearly with pressure. With increase in pressure, permeate flux increases due to availability of enhanced driving force. At higher pressure, higher concentration polarization near the membrane surface causes the increase of membrane surface concentration. Hence, more Lysozyme molecules get transported through the membrane resulting in an increase in permeate concentration and reduction in observed retention. But the effect of osmotic pressure on the permeate flux is negligible due to transmission of most of the Lysozyme molecules (MW: 14600) through the membrane pore (MWCO: 30000). As a result, permeate flux increases almost linearly with pressure. It is observed that with increase in pressure from 220 kPa to 635 kPa, for BSA solution, permeate flux increases from 55 L/m^2 h to 93 L/m^2 h (i.e., 69% increase) and for Lysozyme solution with the same increment of pressure, permeate flux increases from 55.8 L/m^2 h to 135 L/m^2 h (i.e., increase of about 1.4 fold) with decrease in observed retention from 14.5% to 5%.

3.2.3.2. Effect of Solution pH

From the measurement of absorbance at 280 nm for the protein mixture of BSA and Lysozyme (BSA: 0.1 kg/m^3; Lysozyme: 0.1 kg/m^3) in 1mM NaCl solution ant different solution pH, it is observed that with increasing solution pH, the values of absorbance increases and is found to be maximum at a pH around 8.6. On further increase of pH absorbance value decreases. The increase in value of absorbance indicates the aggregation of BSA and Lysozyme molecules. The iso-electric point of BSA is 4.7 and that of Lysozyme is 11.0. Hence BSA and Lysozyme molecules are expected to be negatively and positively charged, respectively within the range of pH under investigation. The strong electrostatic attraction between two oppositely charged protein molecules results in the formation of BSA-Lysozyme complex which is indicated by the appearance of turbidity of the protein solution. With increase in pH from 5.5 to11.0, BSA molecule becomes more negative and Lysozyme becomes less positive (refer Table 3.7). Experimental results confirm that association of BSA and Lysozyme molecules i.e., complex formation is maximum at a pH around 8.6. On further increase in pH, Lysozyme molecule approaches to its iso-electric point. Therefore, electrostatic attraction decreases resulting in a clear solution of protein mixture with low values of absorbance at higher pH. In the present work, the parameters such as protein-protein interaction parameter (B) and the real retention (R_{r2}) are optimized using the algorithm discussed earlier. The optimized values of B' obtained are 329 pa m^6/kg^2, 1117 pa m^6/kg^2, and 324 pa m^6/kg^2 at pH 7.4, 8.6 and 11.0 respectively. The corresponding values of R_{r2} are 0.997, 0.997, and 0.993. Interestingly, at pH 8.6, the higher value of B' indicates the greater extent of electrostatic interaction resulting in the formation of complex between the oppositely charged protein molecules which is consistent with the results of absorbance measurements.

3.2.3.3. Effect of Electric Field on Observed Retention and Permeate Flux

In presence of d.c. electric field, (top plate being anode), electrophoretic movement of free BSA moleclules and complex of BSA and Lysozyme molecules toward the anode resists the particle accumulation on the membrane surface.

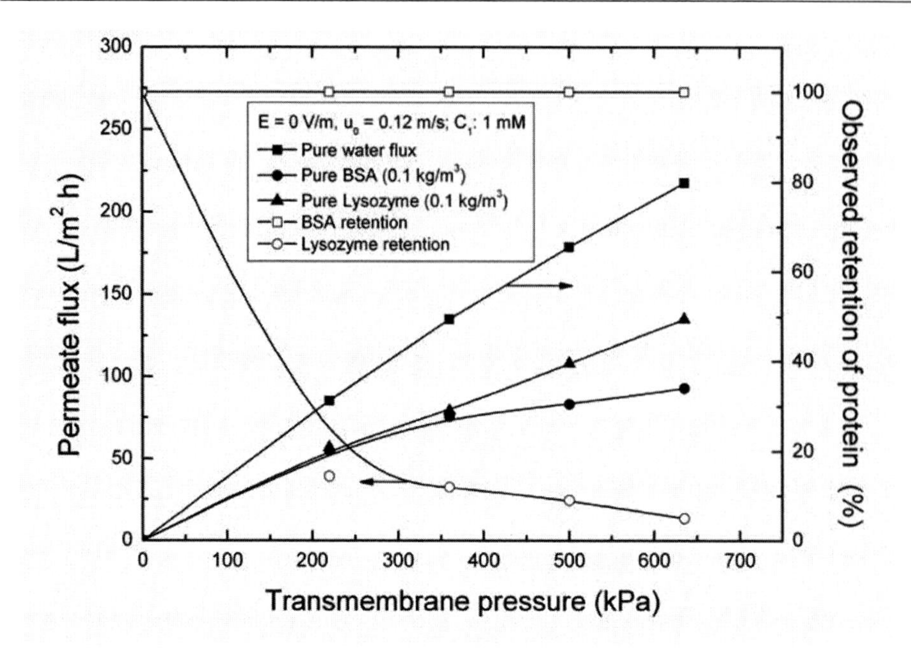

Figure 3.16. Variations of permeate flux and observed retention with transmembrane pressure during ultrafiltration of single protein solution. Open symbols are observed retention of protein and closed symbols are permeate flux.

Therefore concentration polarization decreases with increase in electric field leading to an increase in transmission of Lysozyme molecule by forced convection through the membrane. Experimental results shows that at pH 7.4, a pressure of 500 kPa and a cross flow velocity of 0.12 m/s, for a fixed concentration ratio of BSA and Lysozyme in the feed, with increase in electric field from 0 V/m to 1000 V/m, retention of Lysozyme decreases from 75% to 21%. Model simulated results under predicts the Lysozyme retention in presence of electric field. The retention of Lysozyme is a function of Lysozyme concentration which is less than that in the feed due to complex formation with BSA molecule. On application of electric field with appropriate polarity, the BSA and Lysozyme complex molecules move away from the membrane surface, leaving free Lysozyme whose concentration is less than the Lysozyme concentration in feed whereas in the model the feed concentration of Lysozyme is used as a quantitative estimate of the free Lysozyme concentration can not be made. Therefore, retention corresponds to a value of lower Lysozyme concentration in the feed. This hypothesis is tested by artificially lowering the Lysozyme feed concentration in the model. It is found that at a lower Lysozyme feed concentration of 0.08 kg/m^3, the retention values are close to the experimental values, supporting the understanding of the physics of the process.

When external d.c. electric field is applied with appropriate polarity, the electrophoretic movement of BSA-Lysozyme complex towards the positive electrode causes reduction of concentration polarization at the membrane-solution interface resulting in decrease in membrane surface concentration which in turn decreases the osmotic pressure and hence net driving force increases. This leads to an increase in permeate flux. As a result, permeate flux increases. This can be demonstrated by Figure 3.17. The symbols are experimental data and the solid lines are the model predictions. For example, with an increase in electric field from 0 to 1000 V/m for a fixed feed concentration (BSA: 0.1 kg/m^3, Lysozyme: 0.1 kg/m^3), pressure (500 kPa), cross-flow velocity (0.12 m/s) and pH 7.4, the permeate flux increases from 25.0 L/m^2 h to 38.7 L/m^2 h (about 54.8%), It may also be observed that for a particular electric field, with increase in solution pH from 7.4 to 11.0, permeate flux increases. At pH 7.4, BSA is electro positive and Lysozyme is negative charged. The electrostatic association between the two oppositely charged protein molecules forms a compact layer which tends to deposit on the membrane surface resulting in decrease in permeate flux. In contrast, at pH 11 (which is the isoelectric point of Lysozyme), the electrostatic interaction between BSA and Lysozyme is expected to be minimum leading to higher permeate flux. As discussed earlier with application of electric field, BSA-Lysozyme complex move away from the membrane leading to a reduction in concentration polarization. This results in an increase in permeate flux as well as the transmission of Lysozyme (i.e., decrease in observed retention). As discussed earlier that at pH 7.4, a lowering of Lysozyme concentration to 0.08 kg/m^3, the model fits the experimental data but the effect of lowering Lysozyme concentration in the model does not have significant effect on the permeate flux as shown by the dotted line in Figure 3.17. It can also be seen from the Figure 3.17 that for a fixed concentration ratio of BSA and Lysozyme in the feed, at 1000 V/m, at a pressure of 500 kPa, a cross flow velocity of 0.12 m/s, with increase in solution pH from 7.4 to 11.0, permeate flux increases by a factor of 2.45.

3.2.3.4. Effect of Cross Flow Velocity on Permeate Flux and Observed Retention

Figures 3.18 and 3.19 describe the influence of cross flow velocity on the steady state permeate flux and observed retention for different concentration ratio of BSA and Lysozyme in the feed. These figures compare the experimental results with the model prediction under the same operating conditions. The symbols are the experimental results and the solid lines are the model predictions. It is clear from the Figure 3.18 that for a fixed concentration ratio in the feed, permeate flux increases with increase in cross flow velocity due to its increasing sweeping effect on the membrane surface (less concentration polarization). However, with increase in cross flow velocity, the retention of Lysozyme varies only marginally. Moreover, with increase in concentration of BSA in the feed mixture of BSA and Lysozyme, the membrane surface concentration increases due to enhanced convective transport of BSA-Lysozyme complex towards the membrane surface and consequently osmotic pressure increases and therefore, permeate flux decreases. Since the feed concentration of Lysozyme is kept constant, membrane surface concentration of Lysozyme is almost invariant. But the increase of BSA concentration near the membrane surface leads to a decrease in transmission of Lysozyme through the membrane due to protein-protein interaction and the resistance offered by the BSA-Lysozyme layer on the membrane surface. Hence retention of Lysozyme increases with increase in feed concentration of BSA as observed in Figure 3.19. For example, with increase in BSA concentration from 0.05 kg/m^3 to 0.3 kg/m^3 in the feed, permeate flux decreases from 22.0 L/m^2 h to 17.65 L/m^2 h with increase of Lysozyme retention from 76.5% to 88.5% keeping other operating conditions unchanged (ΔP: 360 kPa, E: 0 V/m, u_0: 0.12 m/s, pH: 8.6). Furthermore, at 0 V/m, the retention of Lysozyme is mostly due to the formation of BSA-Lysozyme complex. Thus the model results are relatively insensitive to initial concentration of Lysozyme and are in good agreement with experimental results. As can be seen from the Figures, a lowering of initial concentration of Lysozyme to 0.08 kg/m^3 in the model, does not affect the permeate flux and retention of Lysozyme (i.e., shown by the dotted lines in Figures 3.18 and 3.19 respectively).

3.2.3.5. Effect of Pressure on Permeate Flux and Observed Retention

The variation of steady state permeate flux and observed retention with operating pressure for different solution pH are described by Figures 3.20 a and b respectively. These figures illustrate the comparison between the experimental data and model prediction under the same operating conditions. The symbols represent the experimental data and the solid lines are the model predictions. It may be seen from the figures that for fixed values of pH and electric field, with increase in pressure, permeate flux increases and retention of Lysozyme decreases. Increase in operating pressure leads to more severe concentration polarization with higher value of Lysozyme concentration on the membrane surface. Increased pressure also facilitates the convective flux through the membrane due to availability of enhanced driving force.

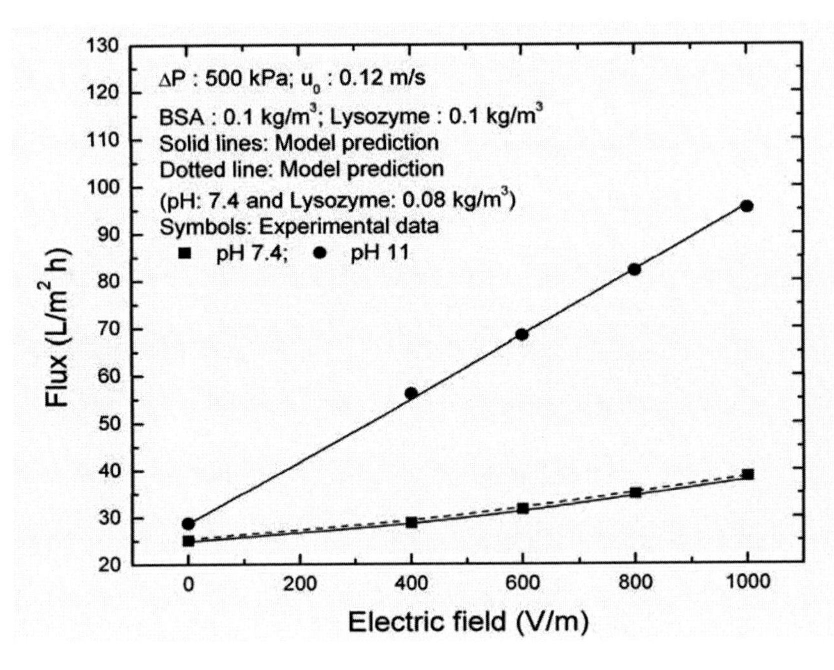

Figure 3.17. Variation of permeate flux with electric field at different solution pH.

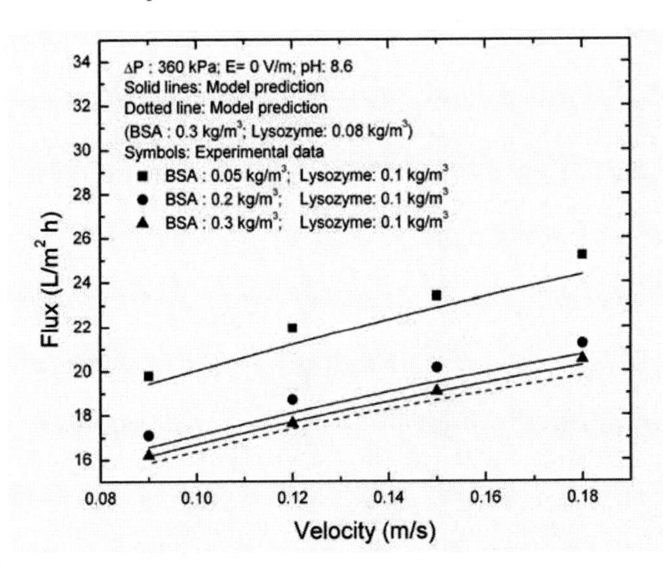

Figure 3.18. Variation of permeate flux with cross flow velocity at different concentration ratio of BSA and Lysozyme in feed mixture.

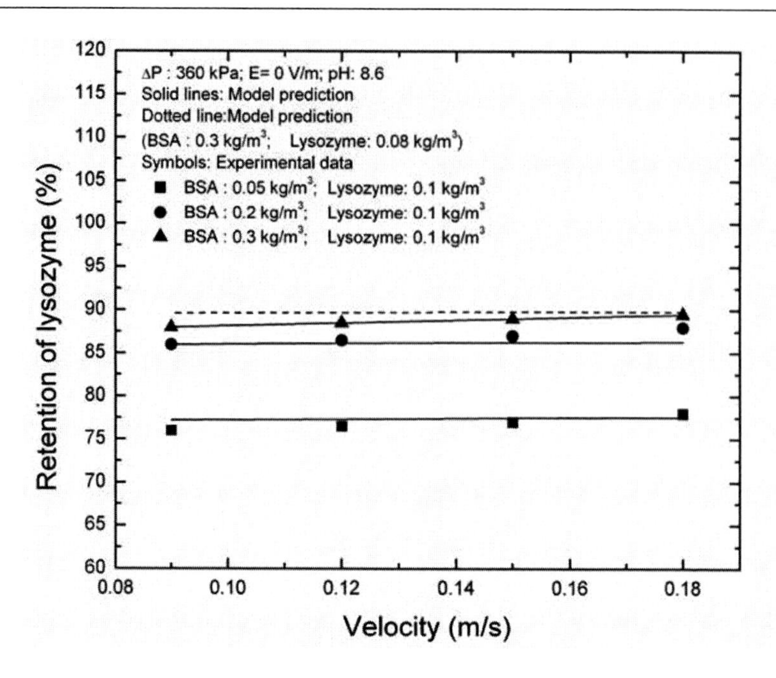

Figure 3.19. Variation of observed retention of Lysozyme with cross flow velocity at different concentration ratio of BSA and Lysozyme in feed mixture.

This leads to an increase in both permeate flux and transmission of Lysozyme molecules through the membrane. Experimental results confirm that for a fixed concentration of BSA and Lysozyme in the feed, at pH 7.4, a cross flow velocity of 0.15 m/s, at 0 V/m, with increase in pressure from 220 kPa to 635 kPa, permeate flux increases from 24.12 L/m^2 h to 27.5 L/m^2 h. (i.e., increase in 14%) and observed retention of Lysozyme decreases from 82% to 73%. Similarly at pH 11.0, with the same increment of pressure, permeate flux increases from 27.7 L/m^2 h to 31.7 L/m^2 h and observed retention of Lysozyme decreases from 60% to 43.5%.

3.2.3.6. Effect of Electric Field on Membrane Surface Concentration

Figure 3.21 represents the plot of membrane surface concentration of both BSA and Lysozyme as a function of electric field for different values of pH keeping pressure, cross flow velocity and concentration ratio of BSA and Lysozyme in the feed unchanged. It is observed from the figure that with increase in electric field, membrane surface concentration of BSA decreases and that of Lysozyme increases.

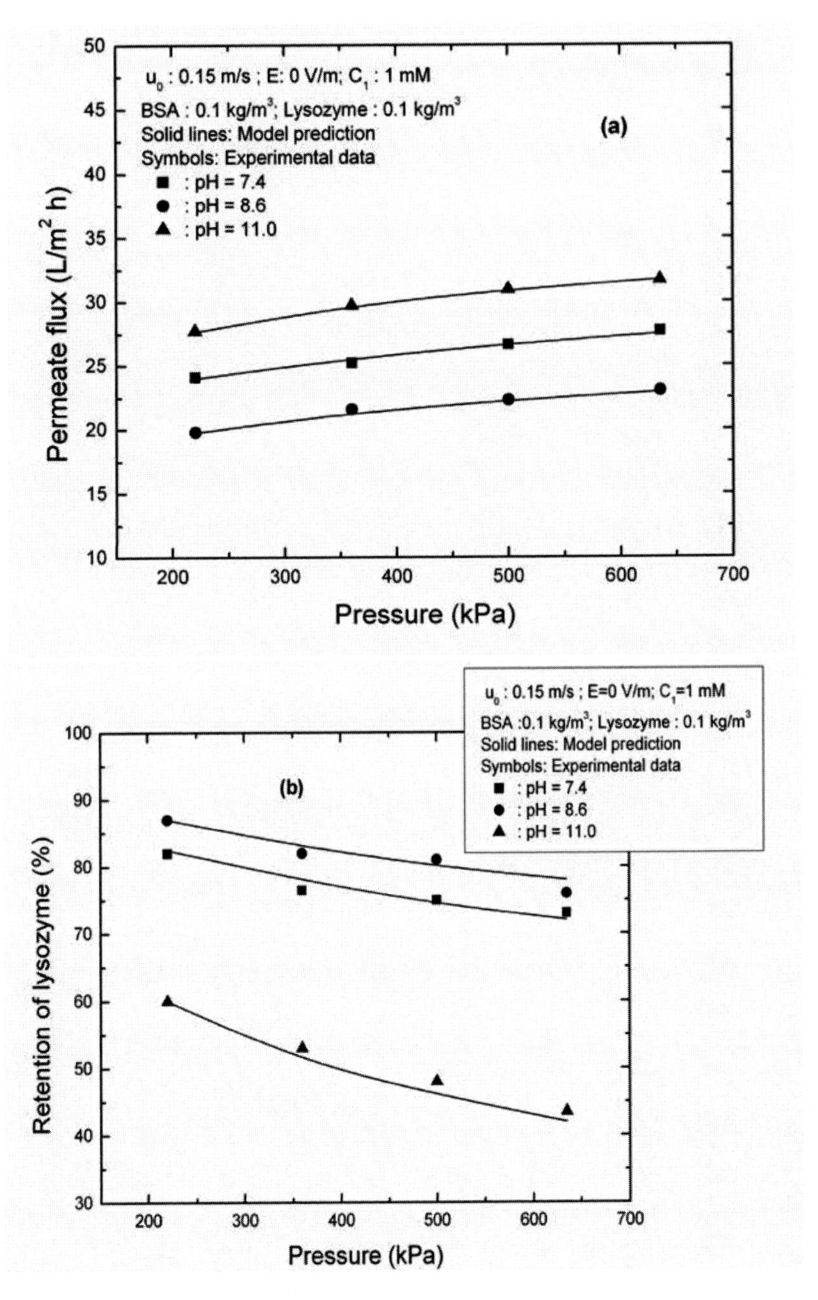

Figure 3.20. Effect of transmembrane pressure on (a) permeate flux (b) observe retention of Lysozyme.

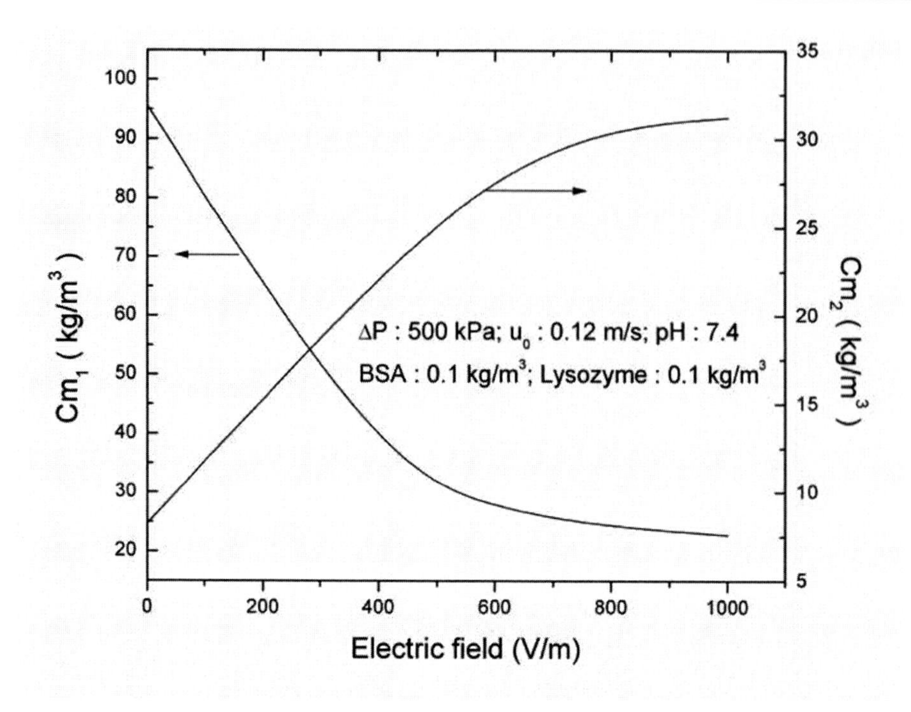

Figure 3.21. Variation of membrane surface concentration of BSA and Lysozyme with electric field.

This is a result of electrophoresis of BSA molecule away from the membrane and that of Lysozyme towards the membrane, assisted further by increased convective flow towards the membrane. For example, at pH 7.4, a pressure of 500 kPa and a cross flow velocity of 0.12 m/s, with increase in electric field from 0 V/m to 1000 V/m, c_{m1} decrease from 95.6 kg/m^3 to 22.8 kg/m^3 whereas, c_{m2} increases from 8.25 kg/m^3 to 31.2 kg/m^3.

3.3. CONCLUSIONS

Application of external d.c. electric field has been found to be quite effective for the separation as well as fractionation of protein solution in terms of permeate flux and process selectivity. A theoretical model incorporating the effects of external d.c. electric field and suction is developed herein for osmotic pressure controlled electro-ultrafiltration process using a completely retentive membrane under laminar flow regime. The model is analytically solved using similarity solution method and the permeate flux values are evaluated under different operating conditions. A Sherwood number relation has been developed to estimate the mass transfer coefficient. Extensive experiments are conducted for the separation of BSA from its aqueous solution under the application of electric field and at different operating conditions, e.g., pressure, flow rate and initial BSA concentrations. It has been observed that application of a suitable d.c. electric field significantly enhances the permeate flux and the values of the mass transfer coefficient For example, for 1.5 kg/m^3 BSA solution, by applying d.c. electric field of 1000 V/m, the permeate flux increases from 34.2 L/m^2 h to 95.4 L/m^2 h compared to that with zero electric field. The predictions from the solutions of the theoretical model are successfully compared with the experimental results. The parametric studies conducted are consistent with the basic understanding of the process. The zeta potentials of membrane determined from streaming potential measurements technique are found to be strongly dependent on pH and ionic strength of the solution. Experimental results confirm that, membrane zeta potential becomes more negative with increasing solution pH and decreasing ionic strength. An increase in solution pH above the iso-electric point of BSA (4.7) increases the electrostatic repulsion between BSA molecules and the membrane, thereby reducing concentration polarization. Thus with increasing solution pH, significant enhancement of permeate flux during separation of BSA solution is observed. The steady state flux is increases from 84.5 L/m^2 hr to 102 L/m^2 hr with increasing pH from 5.5 to 9.0. During fractionation of protein mixture by the pressure driven membrane processes, the major area of concern is the decline in permeate flux and low transmission of the permeable solutes (low molecular weight protein) due to concentration polarization of rejected solutes (mostly high molecular weight protein) and protein-protein interaction. Modeling of polarization phenomena in presence of electric field leading to the prediction of permeate flux and observed retention will be useful for designing of membrane separation process module from both operational and economic point of view. By tuning the operating and physico- chemical conditions, it is possible to attain the most optimum set of operating condition to achieve best fractionation from a systematic parametric study. The detailed parametric study confirms that fractionation increases with increase of electric field and pressure. Similarly permeate flux increases with increase of electric field, pressure and cross flow velocity. It has also been observed that higher permeate flux with maximum separation is achieved when one of the proteins is at its iso-electric point. Experimental results show that at pH 7.4, pressure 500 kPa and cross flow velocity 0.09 m/s, by applying electric field 1000 V/m, permeate flux increases from 25.0 L/m^2 h to 38.7 L/m^2 h and observed retention of Lysozyme decreases from 75% to 21% compared to zero electric field case keeping other operating conditions unchanged. A theoretical model combining film theory for external mass transfer and Darcy's law for the flow through the membrane is proposed. In this model solute-solute interaction and effects of electric field, both are included. The solute-solute interaction parameters are found to be 329 pa m^6/kg^2, 1117 pa m^6/kg^2 and 324 pa m^6/kg^2 for pH 7.4, 8.6 and 11.0 respectively.

APPENDIX 1

Calculation of Osmotic Pressure

Osmotic pressure of a solution containing charged particles is the summation of three interaction forces namely repulsive electrostatic, van der Waals and entropic interactions, which are calculated by using cell model [4].

Electrostatic Interaction

When charged particles interact, their diffused double layers overlap resulting in a repulsive electrostatic force between them. Electrostatic interaction can be calculated from Wigner-Seitz cell model [4] in conjunction with solution of Poisson-Boltzmann equation [2] as,

$$\pi_{electric} = \frac{\sqrt{6}}{A_h} f(d_1) \tag{A.1}$$

$$\text{where, } f(d_1) = \frac{1}{3} S_{cell} n K_B T \left(\mathrm{Cosh} \frac{ze\Phi_{r=r_{cell}}}{K_B T} - 1 \right) \tag{A.2}$$

z is the valency of electrolyte, e is the electronic charge, K_B is the Boltzmann constant, T is temperature, n is the number concentration of electrolyte in the bulk electrolyte and is expressed in terms in terms of molar concentration of electrolytic solution as,

$$n = 1000 \ N_A \ C_1 \tag{A.3}$$

where, c_1 is the ionic strength in molar concentration.

$\Phi_{r=r_{cell}}$ is the electrostatic potential at the cell boundary, S_{cell} ($4\pi r_{cell}^2$) is the surface area of Wigner-Seitz spherical cell of radius, r_{cell}.

$$\text{where, } r_{cell} = (2a + d_1)\left(\frac{3}{4\pi\sqrt{2}}\right)^{\frac{1}{3}} \quad (A.4)$$

A_h is the effective area of a single particle at the hypothetical plane of a regular hexagon drawn about a circle of radius ($a + d_1/2$) and is expressed as,

$$A_h = 2\sqrt{3}\ (a + d_1/2)^2 \quad (A.5)$$

d_1 is the interparticle distance in a hexagonal packing which is related with local volume fraction (φ) as [4,18].

$$d_1 = \left[\left(\frac{4\pi\sqrt{2}}{3\varphi}\right)^{\frac{1}{3}}(a + d) - 2a\right] \quad (A.6)$$

where, φ is the volume fraction in terms of BSA concentration [47],

$$\varphi = \frac{4}{3}\pi a^3 \text{ x } 10^3 \text{ x } N_A \text{ x } \frac{c}{M_w}; \quad (A.7)$$

where, c is concentration in kg/m^3, N_A is the Avogadro's number, a is the effective radius of BSA protein in solution, taken as 2.97nm [18] and d is the distance to surface of shear, taken as 0 .23 nm [18].

Calculation of $\Phi_{r=r_{cell}}$ *in Eq.(A.2)*

The reduced form of electrostatic potential at the cell wall, $\Phi_{r=r_{cell}}$ is,

$$\Psi_{r=r_{cell}} = \frac{ze\Phi_{r=r_{cell}}}{K_BT}$$, which can be calculated as follows:

$$\Psi_{r=r_{cell}} = \frac{A_1}{\beta} e^{\beta-\alpha} + \frac{\xi^1\alpha - A_1}{\beta} e^{-(\beta-\alpha)} \quad \text{(A.8)}$$

where, $A_1 = \dfrac{\xi^1 \alpha e^{-2(\beta-\alpha)}}{\dfrac{\beta-1}{\beta+1} + e^{-2(\beta-\alpha)}}$, $\alpha = \kappa(a+d)$, $\beta = \kappa r_{cell}$

and κ^{-1} is the Debye length, ξ^1 is the reduced zeta potential and is expressed as,

$\xi^1 = \dfrac{ez\xi}{K_B T}$

London-van der Waals Interaction

An attractive van der Waals force is produced during the interaction of two polar molecules. This force can be calculated by differentiating the attractive interaction energy $V_A(d_1)$ with respect to the interparticle distance [4,18,19].

$$F_A(d_1) = \frac{dV_A(d_1)}{dd_1} \quad \text{(A.9)}$$

where,

$$V_A(d_1) = -\frac{A_H}{6}\left(\frac{2a^2}{d_1^2+4ad_1} + \frac{2a^2}{(d_1+2a)^2} + \ln\left(1 - \frac{4a^2}{(d_1+2a)^2}\right)\right) \quad \text{(A.10)}$$

and A_H is the Hamaker constant which depends on the nature of solute particles and the properties of bulk solution. Hamaker constant of BSA has been taken for calculation as 0.753×10^{-20} J [19].

Entropic Pressure

During ultrafiltration membrane surface concentration is higher than bulk concentration. At high concentration, the close packing of the particles generates entropic pressure, which contributes to the osmotic pressure calculation.

The entropic pressure can be calculated by Hall's [48] equation as,

$$\pi_{entropic} = \frac{3\varphi K_B T}{4\pi a^3}\left(\frac{1+\varphi+\varphi^2-0.67825\varphi^3-\varphi^4-0.5\varphi^5-X\varphi^6}{1-3\varphi+3\varphi^2-1.04305\varphi^3}\right) \quad (A.11)$$

where,

$$X = 6.2028 \exp\left(\left(\varphi_{cp}-\varphi\right)\left\{7.9-3.9\left(\varphi_{cp}-\varphi\right)\right\}\right) \quad (A.12)$$

and $\varphi_{cp} = 0.74$

Considering all the above interaction energies osmotic pressure can be calculated as,

$$\pi = \frac{\sqrt{6}}{A_h}\left(f(d_1)+F_A(d_1)\right)+\pi_{entropic} \quad (A.13)$$

The overall osmotic pressure at an ionic strength of 1mM at the different pH and can be expressed in the virial form as follows:

$$\text{For BSA, } \pi_1 = a_1c_1 + a_2c_1^2 + a_3c_1^3 + a_4c_1^4 \quad (A.14)$$

where π_1 is in Pa and c_1 in kg/m^3.

$$\text{For Lysozyme, } \pi_2 = b_1c_2 + b_2c_2^2 + b_3c_2^3 + b_4c_2^4 \quad (A.15)$$

where π_2 is in Pa and c_2 in kg/m^3.

The coefficients Eq.A.14 and A.15 are shown in Table 3.8.

Table 3.8. Values of coefficient of osmotic pressure equation (Eq.A.14 and A.15) of protein solution

Protein	Values of coefficient.			
BSA:	a_1	a_2	a_3	a_4
pH = 7.4	134.70369	- 0.40688	0.00225	-9.475×10^{-7}
pH = 8.6	129.66	- 0.2189	0.0014	3.0×10^{-8}
pH = 11.0	152.00	- 0.2989	0.0015	5.0×10^{-8}
Lysozyme:	b_1	b_2	b_3	b_4
pH = 7.4	33.843	0.0438	0.0002	-4.0×10^{-7}
pH = 8.6	31.302	0.06	0.0001	-4.0×10^{-8}
pH = 11.0	28.186	0.08	7.0×10^{-5}	-3.0×10^{-7}

REFERENCES

[1] W. R. Bowen, P.M. Williams, dynamic ultrafiltration model for proteins: A colloidal interaction approach, *Biotechnol. Bioeng.* 50 (1996) 125-135.

[2] R.J. Hunter, *Zeta potential in colloid Science: Principles and Applications,* Academic Press, London, 1981.

[3] J.N. Israelachvili, *Intermolecular and surface forces,* 2nd Ed. Academic press, London, 1991

[4] W.R. Bowen, F. Jenner, dynamic ultrafiltration model for charged colloidal dispersion: A Wigner-Seitz cell approach, *Chem. Engg. Sci.* 50 (1995) 1707- 1736.

[5] E. Wigner, F. Seitz, On the constitution of metallic sodium, *Phys. rev.* 43 (1933) 804-810.

[6] S.R. Bellara, Z.F. Cui, Maxwell- Stefan approach to modelling the cross flow ultrafiltration of protein solution in tubular membranes, *Chem. Eng. Sci*. 52 (1998) 2153-2166.

[7] S.S. Vasan, R.W. Field, Z. Cui, A Maxwell- Stefan-Gouy-Debye model of the concentration profile of a charged solute in the polarization layer, *Desalination* 192 (2006) 356-363.

[8] A.D. Enevoldsen, E.B. Hansen, G. Jonsson, Electro-ultrafiltraton of industrial enzyme solutions, *J. Membr. Sci.* 299 (2007) 28-37.

[9] A.D. Enevoldsen, E.B. Hansen, G. Jonsson, Electro-ultrafiltraton of amylase enzymes: Process design and economy solutions, *Chem. Eng. Sci.* 62 (2007) 6716-6725.

[10] T. Kappler, C. Posten, Fractionation of proteins with two-sided electro-ultrafiltraton, *Journal of Biotechnology* 128 (2007) 895-907.

[11] P.v. Zumbusch, W. Kulcke, G. Brunner, Use of alternating electrical fields as anti- fouling strategy in ultrafiltration of biological suspensions-Introduction of a new experimental procedure for crossflow filtration, *J. Membr. Sci.* 142 (1998) 75-86.

[12] H. Saveyn, P.V. Meeren, R. Hofmann, W. Stahl, Modelling two-sided electrofiltration of quartz suspensions: Importance of electrochemical reactions, *Chem. Eng. Sci.* 60 (2005) 6768-6779.

[13] G.B. Van der Berg, I.G. Racz, C.A. Smolders, Mass transfer coefficients in cross flow ultrafiltration, *J. Membr. Sci.* 47 (1989) 25-51.

[14] V. Gekas, B. Hallstrom, Mass transfer in the membrane concentration polarization layer under turbulent cross flow. I. Critical literature review and adaptation of existing Sherwood correlations to membrane operations, *J. Membr. Sci.* 30 (1987) 153-170.

[15] S.S. Vasan, C.D. Bain, R.W. Field, Z.F. Cui, A Maxwell–Stefan–Derjaguin– Grahame model of the concentration profile of a charged solute in the polarisation layer, *Desalination* 200 (2006) 175-177.

[16] R.B. Bird, W.E. Stewart, E.N. Lightfoot, *Transport phenomena,* John Wiley, Singapore, 2002.

[17] S. De, S. Bhattacharjee, A. Sharma, P.K. Bhattacharya, Generalized integral and similarity solution of the concentration profiles for osmotic pressure controlled ultrafiltration, *J. Membr. Sci.* 130 (1997) 99-121.

[18] W.R. Bowen, P.M. Williams, The osmotic pressure of electrostatically stabilized colloidal dispersions, *J. Colloid Interface* 184 (1996) 241-250.

[19] W.R. Bowen, A. Mongruel, P.M. Williams, Prediction of the rate of cross-flow membrane ultrafiltration: A colloidal interaction approach, *Chem. Eng. Sci. 51* (18) (1996) 4321-4333.

[20] F.M. White, *Fluid Mechanics,* Mc Graw-Hill, Singapore, 1999.

[21] V.G.J. Rodger, R.E. Spark, Reduction of membrane fouling in ultrafiltration of binary protein mixtures, *AIChE. J.* 37 (1991) 1517-1528.

[22] J.H. Sung, M.S. Chun, H.J. Choi, On the behavior of electrokinetic streaming potential during protein filtration with fully and partially retentive nanopores, *J. Colloid Interface Sci.* 264 (2003) 195-202.

[23] H. Matsumoto, Y. Koyama, A. Tanioka, Interaction of proteins with weak amphoteric charged membrane surfaces: effect of pH, *J. Colloid Interface Sci.* 264 (2003) 82-88.

[24] K.J. Kim, A.G. Fane, M. Nystrom, A. Pihlajamaki, W.R. Bowen, H. Mukhtar, Evaluation of electroosmosis and streaming potential for measurement of electric charges of polymeric membranes, *J. Membr. Sci.* 116 (1996) 149-159.

[25] Y. Soffer, J. Gilron, A. Adin, Streaming potential and SEM-EDX study of UF membranes fouled by colloidal iron, *Desalination* 146 (2002) 115-121.

[26] M.S. Chun, H. Cho, I.K. Song, Electrokinetic behavior of membrane zeta potential during the filtration of colloidal suspensions, *Desalination* 148 (2002) 363-368.

[27] N.D. Lawrence, J.M. Perera, M. Iyer, M.W. Hickey, G.W. Stevens, The use of streaming potential measurements to study the fouling and cleaning of ultrafiltration membranes, *Sep. Purif. Technol.* 48 (2006) 106-112.

[28] L. Ricq, A. Pierre, J.C. Reggiani, J. Pagetti, A. Foissy, Use of electrophoretic mobility and streaming potential measurements to characterize electrokinetic properties of ultrafiltration and microfiltration membranes, Colloid Surf. A: Physicochem. *Eng. Aspects* 138 (1998) 301-308.

[29] P.C. Hiemenz and R. Rajagopalan, *Principles of colloid and surface chemistry,* Marcel Dekker, NY, 1997.

[30] L. Palacio, C.C. Ho, P. Pradanos, A. Hernandez, A.L. Zydney, Fouling with protein mixtures in microfiltration: BSA-lysozyme and BSA- pepsin, *J. Membr. Sci.* 222 (2003) 41-51.

[31] V.L. Vilker, C.K. Colton, K.A. Smith, Concentration polarization in protein ultrafiltration. Part II: Theoretical and experimental study of albumin ultrafiltered in an unstirred cell, *AIChE J.* 27 (1981) 637-645.

[32] R. Van Eijndhoven, S. Saksena, A.L. Zydney,Protein fractionation using electrostatic interaction in membrane filtration, *Biotechnol. Bioeng.* 48 (1995) 406-414.

[33] L. Millesime, J. Dulieu, B. Chaufer, Fractionation of proteins using modified membranes, *Bioseparation* 6 (1996) 135-145.

[34] E. Iritani, Y. Mukai, T. Murase, Upward dead-end ultrafiltration of binary protein mixtures, *Sep. Sci. Technol.* 30 (1995) 369-382.

[35] S. Saksena, A.L. Zydney, Effect of solution pH and ionic strength on separation of albumin from immunoglobulin (IgG) by selective filtration, *Biotechnol. Bioeng.* 43 (1994) 960-968.

[36] N. Ehsani, S. Parkkinen, M. Nyström, Fractionation of natural and model egg- white protein solutions with modified and unmodified polysulfone UF membranes, *J. Membr. Sci.* 123 (1997) 105-119.

[37] K.C. Ingham, T.F. Busby, Y. Sahlestrom, F. Castino, separation of macromolecules by ultrafiltration: Influence of protein adsorption, protein-protein interaction, and concentration polarization, in: A.R. Cooper (Ed.), *Polymer Science and Technology,* vol. 13 (Ultrafiltration Membranes and applications), Plenum Press, New York, 1980.

[38] S. Najarian, B.J. Bellhouse, Effect of liquid pulsation on protein fractionation usingultrafiltration processes, *J. Membr. Sci.*114 (1996) 245-253.

[39] C. Wilharm, V.G.J. Rodgers, Significance of duration and amplitude in transmembrane pressure pulsed ultrafiltration of binary protein mixtures, *J. Membr. Sci.* 121 (1996) 217-228.

[40] M.Y. Teng, S.H. Lin, C.Y. Wu, R.S. Juang, Factors affecting selective rejection of proteins within a binary mixture during cross-flow ultrafiltration, *J. Membr. Sci.* 281 (2006) 103–110.

[41] V.L. Vilker, C.K. Colton, K.A. Smith, The osmotic pressure of concentration protein solutions: effect of concentration and pH in saline solutions of bovine serum albumin, *J. Colloid Interface* 79 (1981) 548-566.

[42] W. Eberstein, Y. Georgalis, W. Saenger, Molecular interaction in crystallizing lysozyme solutions studied by photon correlation spectroscopy, *Journal of Crystal Growth* 143 (1994) 71-78.

[43] M.A. Bos, Z. Shervani, A.C.I. Anusiem, M. Giesbers, W. Norde, J.M. Kleijn, Influence of electric potential of the interface on the adsorption of proteins, *Colloid Surf.* B 3(1994) 91-100.

[44] W.S. Opong, A.L. Zydney, Diffusive and convective protein transport through asymmetric membranes, *AIChE. J.* 37 (1991) 1497-1510.

[45] C.H. Muller, G.P. Agarwal, Th. Melin and Th. Wintgens, Study of ultrafiltration of a single and binary protein solution in a thin spiral channel module, *J. Membr. Sci.* 227 (2003) 51-69.

[46] L. Palacio, C.C. Ho, P. Pradanos, A. Hernandez, A.L. Zydney, Fouling with protein mixtures in microfiltration: BSA-Lysozyme and BSA-pepsin, *J. Membr. Sci.* 222 (2003) 41-51.

[47] V. Karthik, S. DasGupta, S. De, Modeling and simulation of osmotic pressure controlled electro-ultrafiltration in a cross-flow system, *J. Membr. Sci.* 199 (2002) 29-40.

[48] K.R. Hall, Another hard-sphere equation of state, *J. Chem. Phys.* 57 (1972) 2252-2254.

Chapter 4

ELECTRIC FIELD ASSISTED MICELLAR ENHANCED ULTRAFILTRATION

ABSTRACT

This chapter discusses the effect of external d.c. electric field on micellar enhanced ultrafiltration. As a case study, electric field assisted micellar enhanced ultrafiltration (MEUF) has been studied in a rectangular cross flow channel for the removal of dye from aqueous solution over a wide range of operating conditions. In this chapter, the effects of operating conditions e.g., electric field, surfactant to dye concentration ratio, transmembrane pressure and cross flow velocity on the permeate flux and on retention of dye have been discussed. Experimental studies reveal that by applying d.c. electric field (1000 V/m), permeate flux increases by about 25% with approximately 99% removal of dye An expression for the prediction of limiting flux in ultrafiltration is developed.

NOMENCLETURE

a'	Constant used in Eq. (4.6)
b'	Constant used in Eq. (4.6)
c'	Constant used in Eq. (4.7)
c_o	SDS feed concentration, kg/m^3
d_p	Equivalent diameter, m
E	Electric field, V/m

L'	Gel layer thickness, m
L	Channel length, m
k	Constant used in Eq. (4.7)
M_ω	Molecular weight, kg/ k mol
N_A	Avogadro number
R_m	Membrane hydraulic resistance, m^{-1}
R_g	Gel layer resistance, m^{-1}
$v_{w,pure}$	Pure water flux, L/ h. m^2
v_w	Permeate flux, L/ h. m^2
v_w^{limit}	Limiting permeate flux, L/ h. m^2
u	Cross-flow velocity, m/s
x	Parameter used in Eq. (4.12)

Greek Letters

α	Specific gel layer resistance, m/kg
μ	Vviscosity, Pa s
ε_g	Gel porosity, dimensionless
ρ_g	Gel density, kg/m^3
ΔP	Transmembrane pressure drop, kPa
ΔP^{limit}	Limiting transmembrane pressue drop, kPa

4.1. INTRODUCTION

Membrane based separation processes have emerged as an attractive alternative to the conventional separation processes in the large scale treatment of wastewater due to their unique separation capability, easy to scale-up possibilities and low energy consumption. Reverse osmosis (RO) and nanofiltration (NF) are found to be promising methods for high removal of organic molecules [1-4]. The major disadvantages of these processes are membrane fouling, low system throughput, and high pressure requirement. Ultrafiltration (UF) may be a viable alternative in this regard due to its higher throughput at lower operating pressure. However, lower molecular weight organic molecules are not separated by ultrafiltration membrane (molecular weight cut-off 10^3 to 10^6) due to its larger pore size. Such separations are possible using micelle forming surfactant solution. Micellar enhanced ultrafiltration (MEUF), a low energy membrane based separation process appears to be effective in terms of higher selectivity and higher values of permeate flux.

MEUF has been successfully used for the removal of low molecular weight organic molecules [5-13]. In this process, surfactant, at a concentration higher than critical micelle concentration (CMC), is added to the aqueous stream containing ions and organic compounds. Surfactant micelles are found to be electrically charged in aqueous solution. This leads to the formation of electrical double layer (EDL), composed of a stern layer and diffuse layer that restores the electroneutrality in the solution [14]. The excess counter ions around the charged micelle are assumed to be distributed in the double layers. Micelles being charged can adsorb or bind the counter ions onto their surface. Dissolved organics tend to be solubilized in the micelle interior. During ultrafiltration, micelles (having larger size than the membrane pore) containing adsorbed ions and solubilized organics are retained by the membrane. The major drawback encountered in MEUF is the decline in permeate flux due to concentration polarization [15] over the membrane surface and the transmission of the unbound ions, unsolubilized organics and free surfactant monomers with the permeate stream.

Micelles being electrically charged, application of suitable d.c. electric field across the membrane causes the movement of charged micelle away from the membrane surface leading to the reduction of gel-type layer thickness over the membrane surface. Use of electric field has been successfully investigated during cross flow microfiltration of cationic surfactant in aqueous solution for the enhancement of permeate flux [16]. During MEUF, most of the studies are reported regarding the improvement of desired separation. Enhancement of permeate flux during MEUF is also a major concern for the design engineer and currently an area of active research.

Dye Removal: A Case Study

In recent years, with the rapid development of the industries like textile, paper, pulp, plastics, paints, printing, cosmetics etc., the generation of waste water containing synthetic dyes and toxic chemicals increases manifold. Without appropriate treatment, disposal of highly colored and toxic effluent into the environment is harmful to the aquatic and non-aquatic system even at very low concentrations. Moreover, the environmental legislation is increasingly become stringent. Hence to protect environment as well as to keep the higher economic growth rate of the industries, the development of suitable process technology for large scale treatment has become important. Conventional techniques such as oxidation processes [17], flocculation-coagulation [18], biodegradation [19], ozonations [20], and adsorption [21] are generally used for the removal of toxic dye from the effluent. In many cases, these processes are found to be inadequate in removing dyes and organic contaminants due to their stability.

Electric field assisted micellar enhanced ultrafiltration of dye containing effluent is not reported in the literature. This chapter discusses the effect of external d.c. electric field to the enhancement of permeate flux during micellar enhanced ultrafiltration of cationic dye, methylene blue in an aqueous solution in a rectangular cross flow channel over a wide range of operating conditions. The effects of feed surfactant concentration, dye concentration, transmembrane pressure and cross flow velocity on the permeate flux and on observed retention of methylene blue are discussed.

4.2. Principle of Electric Field Assisted MEUF

The cross flow electric field assisted micellar enhanced ultrafiltration (MEUF) system along with development of gel-type layer and external concentration boundary layer of pure surfactant solution and surfactant solution containing dye are schematically presented in Figures 4.1 and 4.2, respectively. During MEUF, surfactant micelles are convected towards the membrane surface by the pressure drop across the membrane. The higher molecular weight micelles are retained by the ultrafiltration membrane, forming a gel-type layer over the membrane surface. Unlike the concentration boundary layer, gel-type layer is usually immobile along the membrane surface. Since the bulk concentration of micelle is much less than that of the gel layer concentration, a concentration boundary layer develops from the bulk of the solution up to the gel layer. This prompts diffusion of gel forming surfactant micelle from the gel layer towards the bulk of the solution due to concentration gradient. In presence of external d.c. electric field (with the top surface maintaining opposite polarity as that of the charged micelle), the charged micelles move away from the membrane surface due to electrophoresis. At steady state, the rate of convective movement of surfactant micelles towards the membrane surface is equal to the rate of migration of surfactant micelles away from the membrane surface due to both back diffusion and electrophoresis.

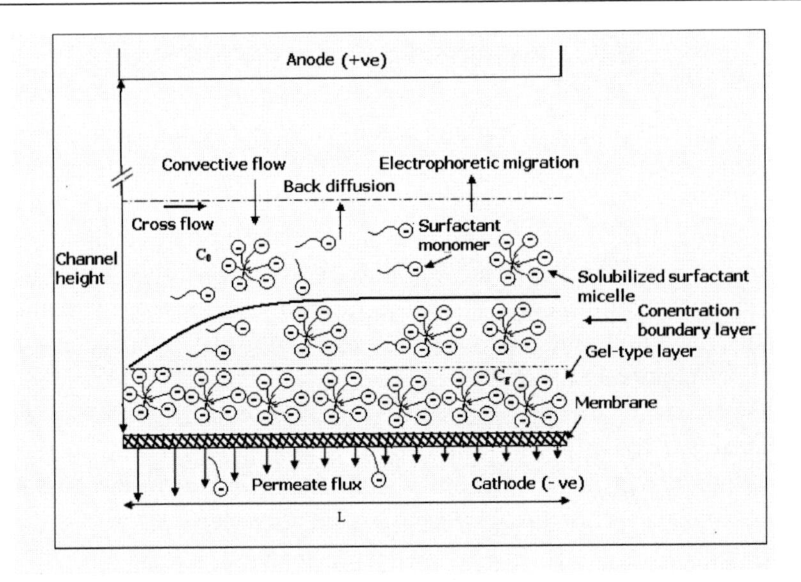

Figure 4.1. Schematic of formation of concentration boundary layer and gel layer by the anionic surfactant micelle over the membrane surface.

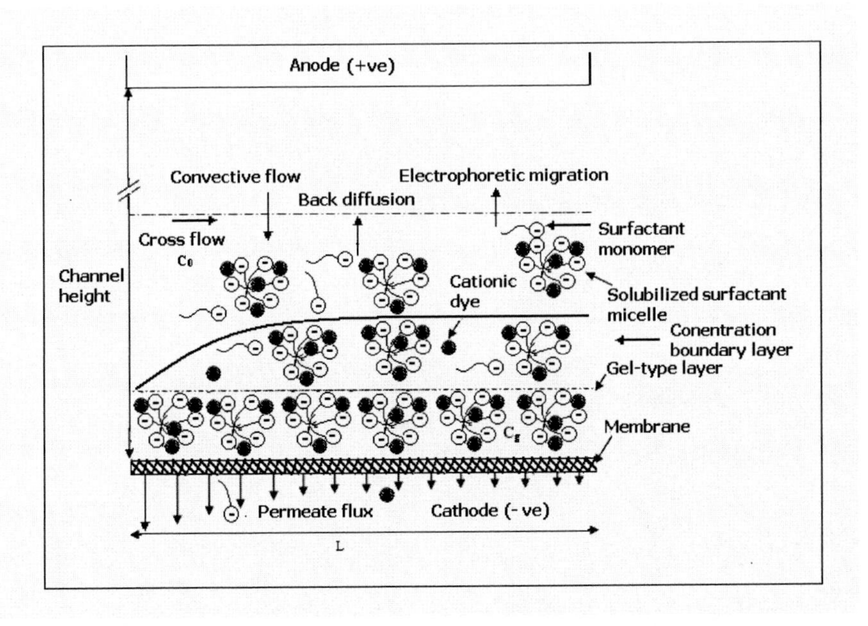

Figure 4.2. Schematic of formation of concentration boundary layer and gel layer by the anionic surfactant micelle containing solubilized dye over the membrane surface.

4.3. LIMITING FLUX PHENOMENA

An important aspect of various ultrafiltration processes is the existence of limiting pressure corresponding to limiting flux i.e., the maximum attainable flux under a given set of operating conditions (i.e., feed concentration and hydrodynamic conditions) [22,23] where an increase in transmembrane pressure does not improve the permeate flux (i.e., existence of flux plateau). Hence performance of ultrafiltration system is restricted by the presence of limiting flux. Moreover, during ultrafiltration, operating pressure is directly associated with the energy consumption of the operation. Hence a prior knowledge about the limiting pressure of energy efficient membrane based process is essential from both operational and design point of view. Quantification of the limiting pressure and limiting flux for a given set of operating conditions (e.g., feed concentration, cross flow velocity and electric field) is discussed in this chapter.

4.3.1. Calculation of Limiting Pressure and Limiting Flux

For pure water, the permeate flux can be expressed as,

$$v_{w,\,pure} = \frac{\Delta P}{\mu R_m} \tag{4.1}$$

During electric field assisted ultrafiltration of a gel forming macromolecular solution, the permeate flux can be expressed by a resistance in series formulation including the flow resistance due to the membrane itself and the deposited gel-type layer.

$$v_w = \frac{v_{w,\,pure}}{\left(1 + \frac{R_g}{R_m}\right)} \tag{4.2}$$

where, R_m and R_g are membrane hydraulic resistance and gel-type layer resistance respectively. Gel-type layer resistance can be characterized using the classical filtration theory. Thus the gel layer resistance (R_g) can be expressed in term of gel layer thickness as,

$$R_g = \alpha\left(1 - \varepsilon_g\right)\rho_g L' \qquad (4.3)$$

where, α is specific gel layer resistance that is obtained from Kozeny-Carman equation as,

$$\alpha = 180\,\frac{\left(1 - \varepsilon_g\right)}{\varepsilon_g^3 d_p^2 \rho_g} \qquad (4.4)$$

where, ε_g, ρ_g and d_p are the porosity, density, and diameter of the gel forming particles respectively. The expression for gel-type layer thickness can be obtained using Eqs. (4.2), (4.3) and (4.4) as follows,

$$L' = \frac{\varepsilon_g^3 d_p^2}{180\left(1 - \varepsilon_g\right)^2}\left[\frac{\Delta P}{\mu v_w} - R_m\right] \qquad (4.5)$$

Knowing the experimental values of permeate flux (v_w), gel porosity (ε_g), diameter of gel forming particle (d_p), membrane hydraulic resistance (R_m), and permeate viscosity (μ), gel layer thickness (L') can be calculated at different operating pressure. The gel layer thickness can then be expressed in terms of transmembrane pressure as follows:

$$L' = a'\,(\Delta P)^{b'} \qquad (4.6)$$

where, a' and b' are constant. Using Eqs. (4.3), (4.4) and (4.6), the expression of $\frac{R_g}{R_m}$ becomes

$$\frac{R_g}{R_m} = c'\,(\Delta P)^{b'} \qquad (4.7)$$

where, $c' = 180\,\dfrac{\left(1 - \varepsilon_g\right)^2 a'}{\varepsilon_g^3\, d_p^2\, R_m}$

Therefore, using Eq. (4.7), the permeate flux can be obtained from Eq. (4.2) as,

$$v_w = \frac{\Delta P}{\mu R_m \left[\left(1 + c' (\Delta P)^{b'}\right)\right]} \tag{4.8}$$

In order to observe the effect of pressure on permeate flux, Eq. (4.8) is differentiated with respect to ΔP as,

$$\frac{dv_w}{d(\Delta P)} = \frac{1}{\left[\mu R_m \left(1 + c' (\Delta P)^{b'}\right)\right]} - \frac{c' b' \left(\Delta P\right)^{b'}}{\left[\mu R_m \left(1 + c' (\Delta P)^{b'}\right)\right]^2} \tag{4.9}$$

For limiting flux condition, $\frac{d v_w}{\Delta P} = 0$ and from Eq. (4.9) one can get an expression for ΔP^{limit} as,

$$\Delta P^{limit} = \left[\frac{1}{c' \left(b'-1\right)}\right]^{\frac{1}{b'}} \tag{4.10}$$

and the expression for limiting flux:

$$v_w^{limit} = \frac{\Delta P^{limit}}{\left[\mu R_m \left(1 + c' (\Delta P^{limit})^{b'}\right)\right]} \tag{4.11}$$

4.4. Membrane Experiments

4.4.1. Chemicals

All chemicals are of reagent grade. Anionic surfactant, sodium dodecyl sulfate, SDS (MW 288.38), chloroform ($CHCl_3$, MW 119) are obtained from M/s, Merck Ltd., Mumbai, India. The cationic dye, methylene blue, MB (MW 373.90) is purchased from Aldrich Chemical Company, Inc. Hyamine 1622 (0.004 M)($C_{27}H_{42}NO_2Cl$, FW 448) and disulfine blue VN ($C_{27}H_{31}N_2NaO_6S_2$, MW 567) is obtained from Merck KGaA, 64271 Darmstadt, Germany and dimindium bromide ($C_{20}H_{18}BrN_3$, MW 380) is procured from Loba Chemie Pvt. Ltd, Mumbai, India. Critical micelle concentration of SDS surfactant is

2.32 kg/m^3 [24]. The surfactant is used as, supplied, without any further treatment. All the feed solutions are prepared using doubled distilled water.

4.4.2. Membrane

An ultrafiltration membrane (Polyphenylene ethersulfone) of molecular weight cut-off (MWCO) 10 kDa, obtained from M/s, Permionics Membranes Pvt. Ltd., Boroda, Gujrat, (India), has been used for all the MEUF experiments. The membrane is hydrophilic, compatible in a pH range 2-12. Membrane permeability is measured using distilled water and is found to be 5.5×10^{-11} m/Pa s.

4.4.3. Cross Flow Cell and Operating Conditions

Details of cross flow cell have been discussed in Chapter 2 (see section 2.3.1). Electric field assisted micellar enhanced ultrafiltration experiments are designed to observe the effects of operating conditions (electric field, transmembrane pressure, SDS concentration, MB concentration and cross flow velocity) on the steady state permeate flux and retention. During experiments, one parameter is varied while other three are kept unchanged to get the exact picture of dependence. For the experiments of pure SDS in aqueous solution, the concentrations of SDS are selected as 5, 15, 20, 25 and 30 kg/m^3; electric fields are chosen as 0, 400, 600, 800 and 1000 V/m; transmembrane pressures are varied as 220, 360, 500 and 635 kPa; cross flow velocities are selected as 0.09, 0.12, 0.15, 0.18 m/s. For the experiments of SDS and methylene blue (MB) mixture, the concentrations of MB dye are chosen as 0.01, 0.02, 0.03, 0.04 and 0.05 kg/m^3. Other variables remain unaltered as that of pure surfactant experiments.

4.4.4. The Steps Used in Experiment

During cross flow experiments, membrane compaction is carried out at a pressure of 690 kPa (higher than the maximum operating pressure) for 3 hours using distilled water. During membrane compaction, the flux values are continuously measured until constant flux is obtained. The membrane permeability is obtained from the slope of steady state flux versus transmembrane pressure drop. For each MEUF experiments, feed solution is prepared by dissolving required amount of surfactant and dye in doubled distilled water. The solution is kept under slow stirring for about 15 minutes and then kept static for around 2 hours. The feed is then pumped to the cell. The permeate is collected from the bottom of the cell. The duration of each experiment is around 40-45 minutes. After each experiment, the membrane is thoroughly washed, in situ, by distilled water for thirty minutes. The cross flow channel is dismantled thereafter, and the membrane is dipped in distilled water for 30 minutes. It is then carefully washed repeatedly with distilled water to remove trace amount of surfactant. The cross flow cell is reassembled and the membrane permeability is checked again. It is seen that the membrane permeability remains almost constant between successive runs. All the experiments are carried out at room temperature ($25 \pm 2^{\circ}$C.). Experiments are repeated thrice for all operating conditions to check its repeatability.

4.4.5. Analysis of Feed and Permeate

4.4.5.1. Concentration of SDS and Methylene Blue

SDS concentration is determined by a two-phase titration [25]. The titrant is benzothonium chloride (hyamine 1622), a cationic surfactant, the indicator is an acidic mixture of a cationic dye (dimindium bromide) and an anionic dye (disulfine blue VN). The titration is carried out in a water chloroform medium. SDS concentration is determined using the following equation:

$$\text{SDS concentration} = \frac{x \times \text{molar concentration of hyamine} \times \text{M.W. of SDS monomer}}{5 \text{ ml of sample}} \tag{4.12}$$

where, x is the volume (ml) of hyamine 1622 required for titration.

The concentration of methylene blue (maximum wavelength 665 nm) in the feed and permeate stream is measured by a UV spectrophotometer (GENESYS 2, Thermo Spectronic, USA). The effect of surfactant on the absorbance of MB is found to be almost negligible.

4.4.5.2. Measurement of Viscosity, Conductivity and pH

Viscosity, conductivity and pH of the solutions are measured by Ostwald capillary viscometer, autoranging conductivity meter (Toshniwal Instrument, India), Orion Aplus™ Benchtop pH Meter (M/s, Thermo Electron Corporation, Beverly, MA, U.S.A.), respectively, at 25±2°C. For the operating feed concentration of SDS in aqueous solution (i.e., from 5 kg/m^3 to 30 kg/m^3), pH and conductivity of the solution are found to vary from 8.6 to 9.6 and from 0.75 mS/cm to 3.1 mS/cm respectively. Viscosity of the solutions varies from $0.95x10^{-3}$ Pa s to $1.03x10^{-3}$ Pa s.

4.4.6. Determination of Particle Size

The particle size and zeta potential of the micelles are determined by Malvern Zetasizer (Nano ZS, ZEN3600, Malvern, England). It measures the particle size using the technique of dynamic light scattering (DLS). The suspended particles are in brownian motion. DLS measures the brownian motion of the particle to calculate the size of the particle. In this system Helium-Neon red laser is used to provide the source of light at a wavelength of 638.2 nm and a detector is used to measure the intensity of scattered light at an angle of 173°, called backscattered detection. The rate of intensity fluctuation is used to calculate the size of the particle. Thus DLS gives a measure of intensity particle size distribution shown in Figure 4.3. The average particle size calculated from this distribution is found to be 3.83 nm which agree well with the reported value [26]. This value is found to be almost invariant with the initial concentration of the surfactant. Zeta potential is obtained by measuring electrophoretic mobility and then by using Henry equation. The electrophoretic mobility of the particle is measured by Laser Doppler Velocimetry. Zeta potential of charged particle is a function of pH and ionic strength. The value of zeta potential of SDS micelle is found to be in the range from -70 mV to -77 mV for the SDS solution under the present operating conditions. Zeta potential of SDS micelle is found to be -71 mV at a pH of 9.4 and at a SDS concentration of 20 kg/m^3.

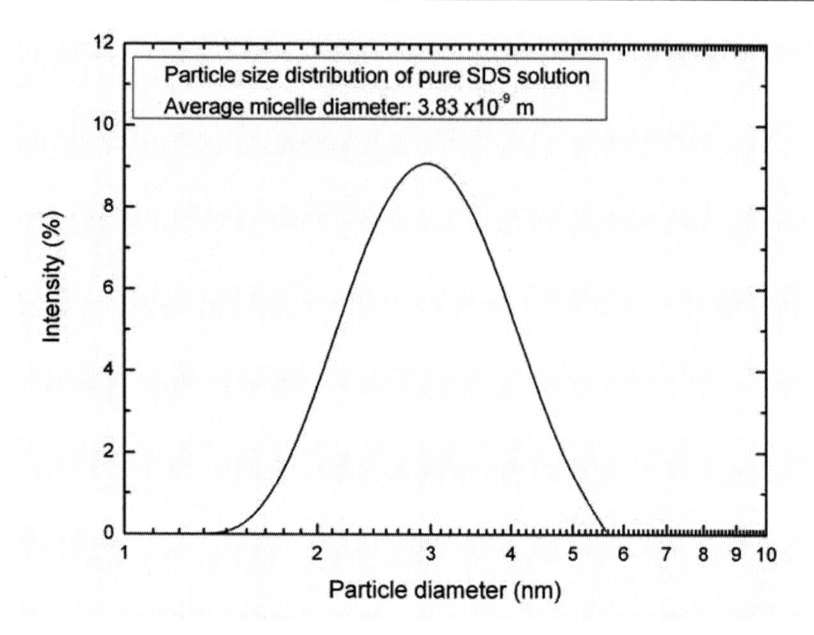

Figure 4.3. Particle size distribution of SDS solution at a concentration higher than its CMC value.

4.4.7. Determination of Gel Porosity

The gel-type layer of SDS micelle is assumed to be expressed in terms of particle diameter and gel concentration by the following expression [27]

$$\varepsilon_g = 1.0 - 10^3 \frac{\pi}{6} c_g \frac{N_A d_p^3}{M_w} \tag{4.13}$$

where, N_A is the Avogadro number, M_w is the molecular weight of the micelle, c_g is the micelle gel concentration which is taken as 164 kg/m^3 [24], and d_p is the diameter of SDS micelle. Assuming the aggregation number to be 64 [28], the molecular weight of SDS micelle becomes 18456. Thus the porosity of SDS micelle gel layer is found to be 0.8426 from Eq. (4.13).

4.5. ANALYSIS OF PERMEATE FLUX, GEL-LAYER THICKNESS AND SOLUTE RETENTION

4.5.1. Electric Field Assisted MEUF of Pure Surfactant Solution

4.5.1.1. Variation of Gel Layer Thickness with Pressure

Variation of gel layer thickness with transmembrane pressure for pure SDS solution, as obtained from model calculation (Eq. 4.5), is shown in Figure 4.4, for with and without electric field cases. It is observed from the figure that with increasing pressure (ΔP), gel layer thickness (L') on the membrane surface increases almost linearly. Similar trends are observed for with and without electric field cases. This is due to the fact with increase in pressure, more SDS micelles are convected towards the membrane surface resulting in an increase in gel layer thickness.

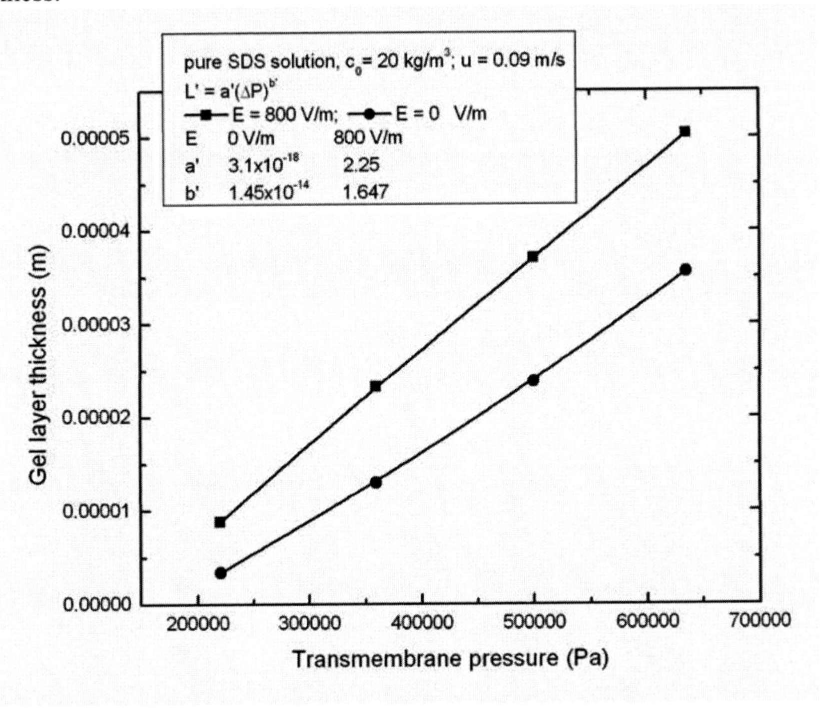

Figure 4.4. Variation of gel layer thickness with transmembrane pressure for pure SDS solution at a cross flow velocity of 0.09 m/s for E = 0 V/m and E = 800 V/m.

The correlation between gel layer thickness and transmembrane pressure is obtained from curve fitting and can be written as follows:

$L' = a'(\Delta P)^{b'}$. In this calculation, L' is in meter and ΔP is in Pa. Knowing the values of coefficients a' and b' the values of limiting pressure and limiting flux can be calculated (using Eqs. 4.10 and 4.11) for a given set of operating conditions. The values of limiting pressure and limiting flux for various operating conditions are depicted in Table 4.1.

Table 4.1. Limiting pressure and limiting flux during ultrafiltration of pure SDS solution at various operating conditions

c_o (kg/m^3)	u = 0.09 m/s, E = 0 V/m				u = 0.15 m/s, E = 0 V/m			
	a'	b'	ΔP^{limit} (kPa)	V_w^{limit} (L/m^2 h)	a'	b'	ΔP^{limit} (kPa)	V_w^{limit} (L/m^2 h)
15	5.05E-15	1.716	668	55.08	6.26E-16	1.857	675	61.63
20	1.45E-14	1.647	657	51.12	8.92E-16	1.835	663	59.76
25	1.76E-13	1.484	648	41.76	2.34E-14	1.615	654	49.32
30	5.21E-13	1.417	630	36.65	1.78E-13	1.485	635	42.12

c_o (kg/m^3)	u = 0.09 m/s, E = 800 V/m				u = 0.15 m/s, E = 800 V/m			
	a'	b'	ΔP^{limit} (kPa)	V_w^{limit} (L/m^2 h)	a	b	ΔP^{limit} (kPa)	V_w^{limit} (L/m^2 h)
15	3.11E-19	2.413	580	67.32	2.76E-20	2.570	637	77.04
20	3.1E-18	2.250	579	63.36	4.26E-19	2.378	626	71.64
25	6.19E-16	1.879	572	53.28	1.53E-17	2.139	570	60.00
30	8.92E-15	1.697	565	45.94	4.83E-16	1.900	556	51.84

4.5.1.2. Effect of Pressure on Permeate Flux

In Absence of Electric Field

For various feed concentrations of pure SDS micellar aqueous solutions, the variation of steady state permeate flux with transmembrane pressure are shown in Figure 4.5 (a and b) The symbols are experimental data and solid lines are the calculated results obtained from Eq. 4.6. From the figure it is clear that at lower operating pressure, permeate flux increases almost linearly for lower and gradually for higher operating pressure values. Surfactant micelles form a gel-type layer over the membrane surface [29]. It is clear from the Figure 4.5a and 4.5b that at lower operating pressure, gel layer thickness is weakly dependent on pressure, resulting in almost linear increase of permeate flux with driving force (i.e., transmembrane pressure). At higher operating pressure, gel layer thickness varies strongly with pressure leading to an increase in gel layer resistance which compensates the enhanced pressure drop. The pressure at which the nature of curve changes from linear to non-linear is called critical pressure [30]. Beyond this, a pressure independent flux i.e., limiting flux is obtained and the corresponding pressure in called limiting pressure. For example, at zero electric field and at a SDS feed concentration of 20 kg/m^3, the limiting pressures are found to be 657 kPa and 663 kPa for cross flow velocities of 0.09 m/s and 0.15 m/s, respectively. The corresponding values of limiting fluxes are 51.12 L/m^2 h and 59.76 L/m^2 h, respectively. The values of limiting pressure for various operating conditions are tabulated in Table 4.1. These data provide us important information regarding the maximum operating pressure (just below the limiting pressure) that could be used for such systems. It is also observed from the figure that the permeate flux decreases with feed concentration. As feed concentration increases, more SDS micelles are convected towards the membrane leading to increase in thickness of gel type layer. Hence permeate flux decrease with increases in feed concentration. The effects of feed concentration on the limiting pressure are not significant as observed from experimental results. For example, at zero electric field, with increase in SDS feed concentration from 15 kg/m^3 to 30 kg/m^3, the values of limiting pressure decreases from 668 kPa to 630 kPa and 675 kPa to 635 kPa, for cross flow velocities of 0.09 m/s and 0.15 m/s, respectively.

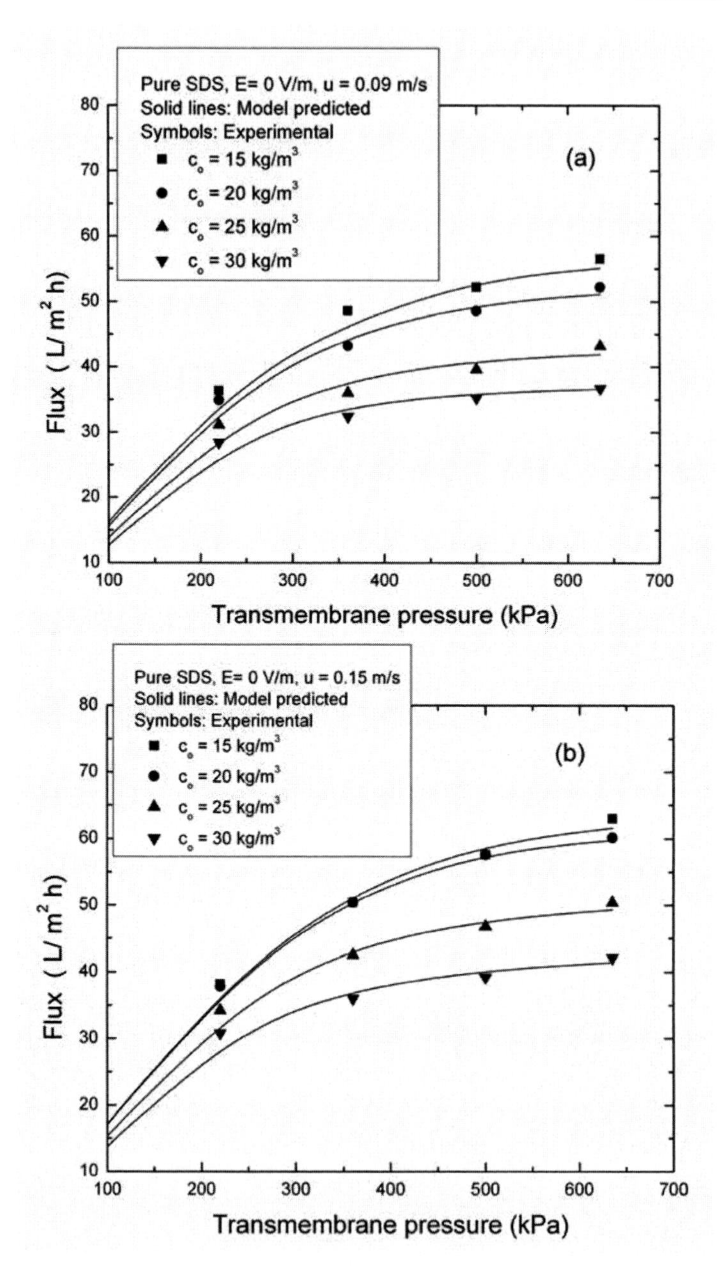

Figure 4.5. Variation of permeate flux with transmembrane pressure for different SDS concentration at zero electric field and (a) at 0.09 m/s and (b) at 0.15 m/s.

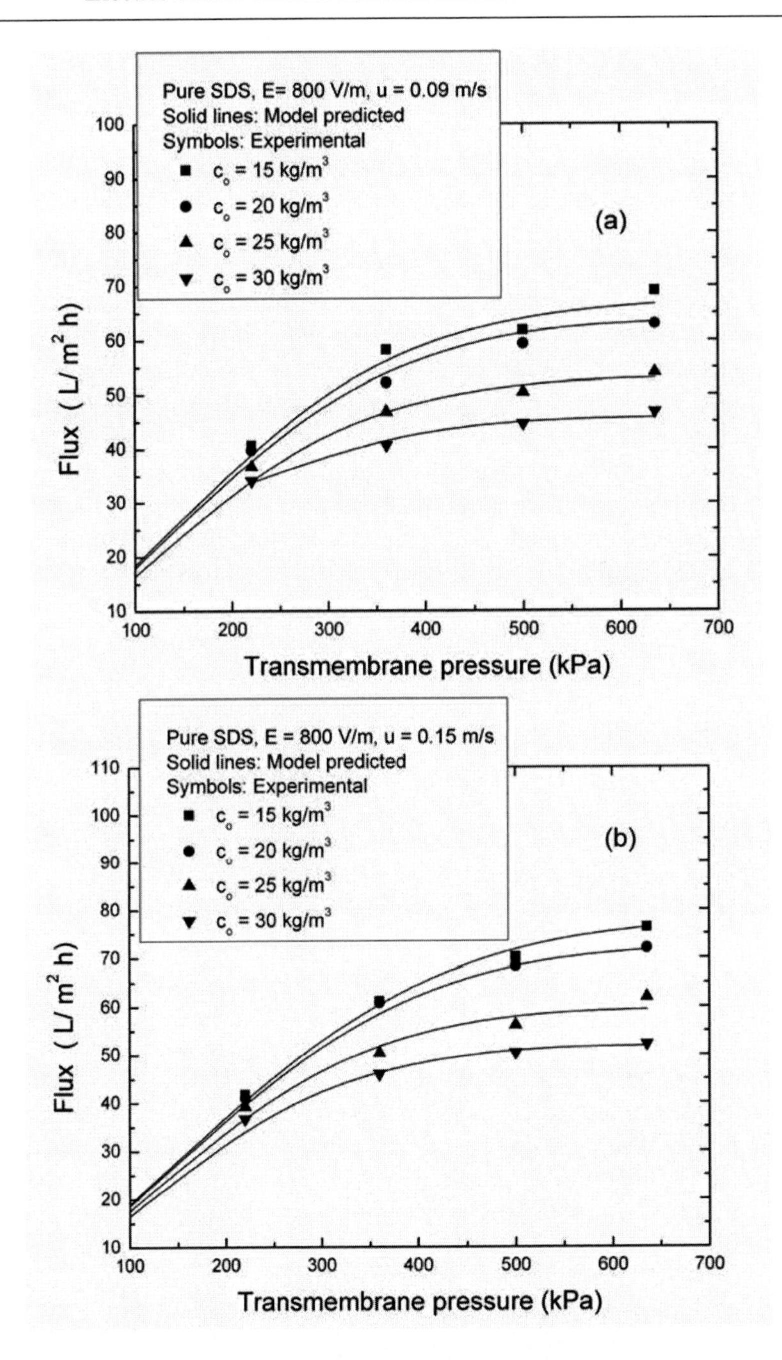

Figure 4.6. Variation of permeate flux with transmembrane pressure for different SDS concentration at 800 V/m and (a) at 0.09 m/s and (b) at 0.15 m/s.

In Presence of Electric Field

Figure 4.6(a and b) demonstrates the variation of steady state permeate flux with transmembrane pressure for various feed concentrations of pure SDS micellar solutions in presence of electric field. The symbols are experimental data and solid lines are calculated results. In presence of electric field, with increase in pressure, permeate flux follows the similar trends as observed in case of zero electric field case for reasons already discussed earlier. It is further noticed that in presence of electric field, higher values of permeate fluxes are obtained compared to no electric field case (Figure 4.5). For example, at 20 kg/m^3 feed concentration, 360 kPa pressure and 0.09 m/s cross flow velocity, the flux enhances from 43.1 L/m^2 h to 52.2 L/m^2 h, at 800 V/m electric field resulting in 21.0% increase in permeate flux. This phenomenon is due to migration of charged SDS micelles away from the membrane surface by electrophoresis resulting in decrease in the gel layer thickness and hence, gel layer resistance. For example, at 800 V/m and at a SDS feed concentration of 20 kg/m^3, the limiting pressures are found to be 579 kPa and 626 kPa for cross flow velocities of 0.09 m/s and 0.15 m/s, respectively. The corresponding values of limiting fluxes are 63.36 L/m^2 h and 71.64 L/m^2 h, respectively. Interestingly, it may also be observed that the values of limiting pressure under electric field are found to be less compared to without electric field (as presented in Table 4.1). Therefore, in presence of electric field, a significant flux augmentation is achievable with a lower operating pressure, thereby, decreasing the process operating cost. In presence of electric field, with increase in feed concentration limiting pressure decreases. For example, at 800 V/m, with increase in SDS feed concentration from 15 kg/m^3 to 30 kg/m^3, the values of limiting pressure decreases from 580 kPa to 565 kPa and 637 kPa to 556 kPa, for cross flow velocities of 0.09 m/s and 0.15 m/s, respectively.

4.5.1.3. Effect of Pressure on Gel Layer Thickness

Figure 4.7 (a and b) show the variation of gel layer thickness and non-dimensional resistance (R_g/R_m) with transmembrane pressure for different feed concentration of pure SDS solution at zero electric field. Closed symbols are for gel layer thickness and open symbols are for the resistance. It is observed from both the figures that for a fixed feed concentration both gel layer thickness and non-dimensional resistance increase with increase in pressure. The experimental results confirm that for a SDS concentration of 20 kg/m^3, at 0.09 m/s, with increase in pressure from 220 kPa to 635 kPa, both gel layer thickness and non-dimensional resistance increase by a factor of about 6 and this factor is about 7 for 0.15 m/s. Furthermore, gel layer thickness is a strong function of feed concentration. For example, a pressure of 360 kPa, with increase in feed concentration from 15 kg/m^3 to 30 kg/m^3, gel layer thickness increases from 20.8

μm. to 43.0 μm and from 14.8 μm. to 35.0 μm for cross flow velocities 0.09 m/s and 0.15 m/s, respectively.

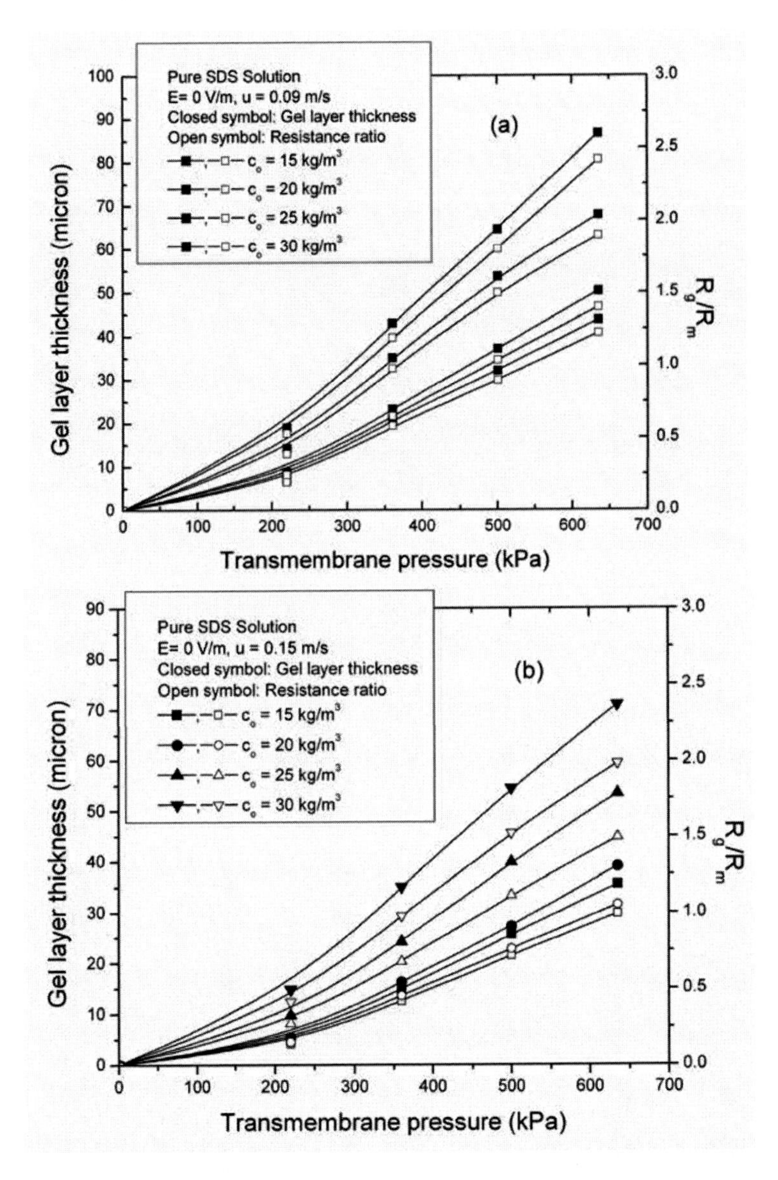

Figure 4.7. Variation of gel layer thickness and ratio of gel resistance to membrane resistance with transmembrane pressure for different SDS concentration at zero electric field and (a) at 0.09 m/s and (b) at 0.15 m/s.

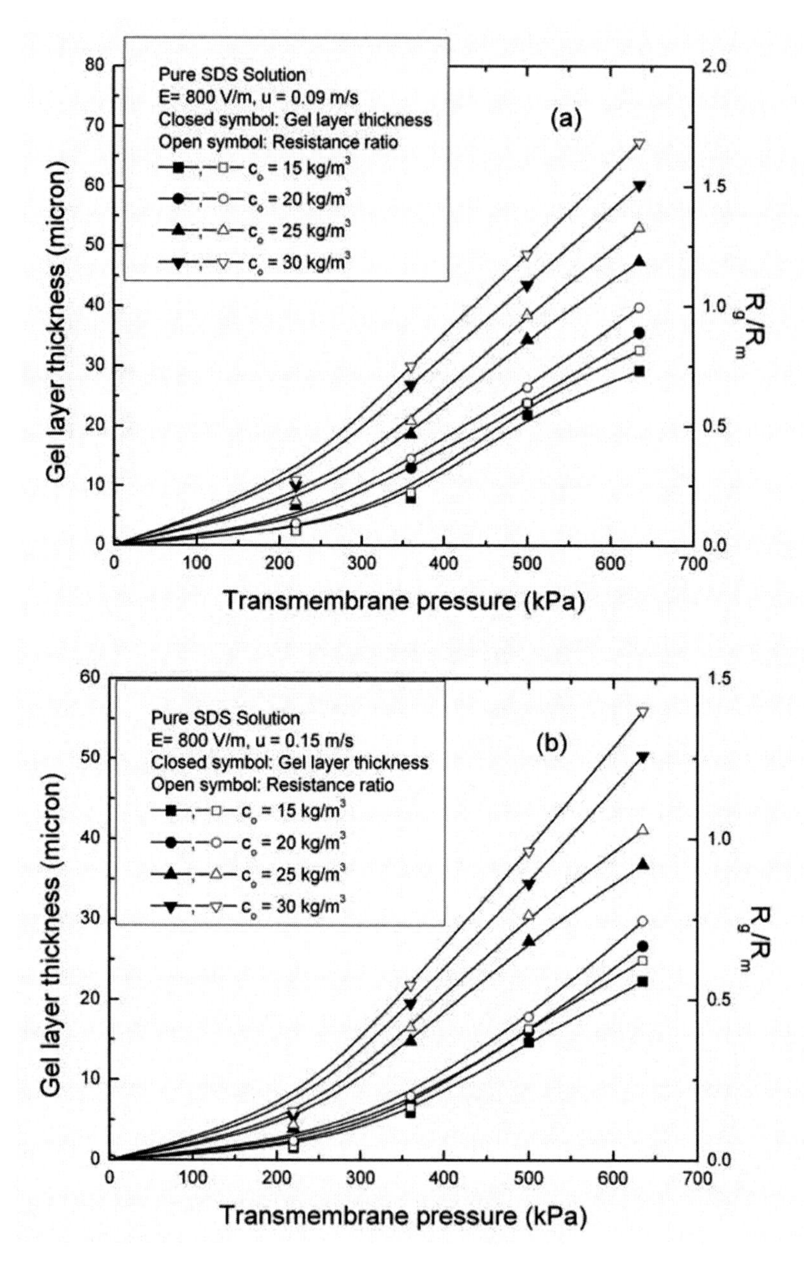

Figure 4.8. Variation of gel layer thickness and ratio of gel resistance to membrane resistance with transmembrane pressure for different SDS concentration at 800 V/m and (a) at 0.09 m/s and (b) at 0.15 m/s.

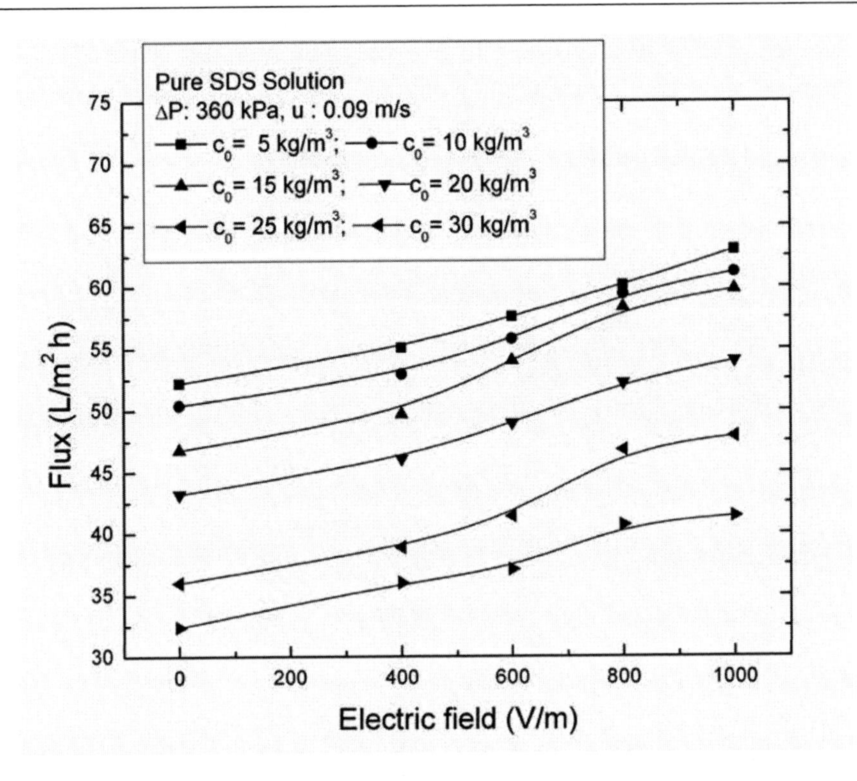

Figure 4.9. Variation of permeate flux with electric field for different SDS concentration at a fixed pressure of 360 kPa and a cross flow velocity of 0.09 m/s.

Similar trends are also observed in the presence of electric field shown in Figure 4.8 (a and b). For a fixed feed concentration of 20 kg/m^3 SDS solution and a cross flow velocity of 0.09 m/s and an electric field of 800 V/m, with increase in pressure from 220 kPa to 635 kPa both gel layer thickness and dimensionless resistance increase by a factor of about 11, whereas, at 0.15 m/s this factor is about 10. Furthermore, as observed in case of zero electric field, gel layer thickness shows a strong dependence on feed concentration. For example, a pressure of 360 kPa, with increase in feed concentration from 15 kg/m^3 to 30 kg/m^3, gel layer thickness increases from 8 μm. to 27 μm and from 6 μm. to 20 μm for cross flow velocities 0.09 m/s and 0.15 m/s, respectively.

4.5.1.4. Effect of Electric Field on Permeate Flux

Figure 4.9 shows the variation of permeate flux with electric field for different feed concentration of pure SDS solution. It may be observed from the figure that for a fixed feed concentration with increase in electric field permeate flux increases as discussed earlier. In presence of external d.c. electric field with appropriate polarity, electrophoretic migration of charged SDS micelles towards the positive electrode causes reduction of deposition on the membrane surface and gel layer thickness is restricted leading to enhancement of permeate flux. It has also been observed that effect of electric field is slightly higher at higher SDS concentration. For example, for a fixed feed concentration of 5 kg/m^3 SDS solution, pressure of 360 kPa and cross-flow velocity of 0.09 m/s, with an increase in electric field from 0 to 1000 V/m, the permeate flux increases from 52.2 L/m^2 h to 63.0 L/m^2 h resulting in 20.7% enhancement in permeate flux. whereas, for 30 kg/m^3 SDS solution, flux enhances from 32.4 L/m^2 h to 41.4 L/m^2 h (i.e., 27.7% enhancement) keeping other operating conditions unchanged.

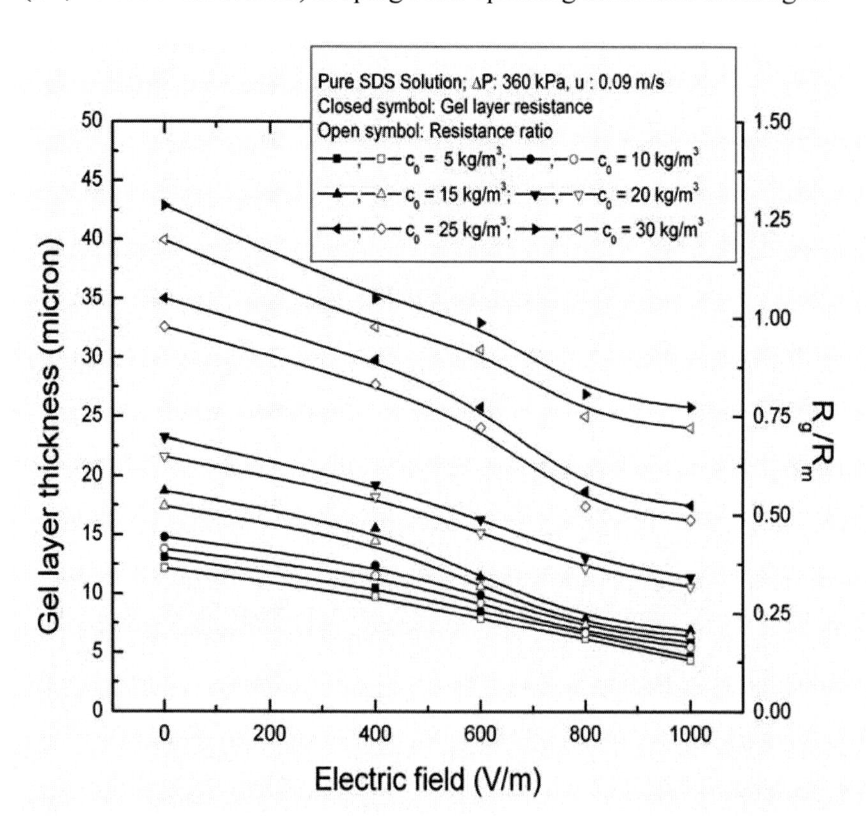

Figure 4.10. Variation of gel layer thickness and ratio of gel resistance to membrane resistance with electric field for different SDS concentration at a transmembrane pressure of 360 kPa and at a cross flow velocity of 0.09 m/s.

4.5.1.5. Effect of Electric Field on Gel Layer Thickness

The variation of gel layer thickness and non-dimensional resistance with electric field for different feed concentration of pure SDS solution is illustrated in Figure 4.10. It is observed from the figure that for a fixed feed concentration with increase in electric field both gel layer thickness and non-dimensional resistance decreases. At higher electric field, under the same operating conditions, the concentration of SDS micelles near the membrane surface is lower due to electrophoretic migration of gel forming SDS micelles away from the membrane surface resulting in a decrease in gel layer thickness and hence gel layer resistance decreases. For example, for a fixed SDS concentration of 20 kg/m^3, at a pressure of 360 kPa and a cross flow velocity of 0.09 m/s, with an increase of electric field from 0 to 1000 V/m, both gel layer thickness and (R_g/R_m) become almost half. As observed from the figure both the values of gel layer thickness and (R_g/R_m) increases with feed concentration keeping other operating conditions unaltered. For example, at a pressure of 360 kPa, a cross flow velocity of 0.09 m/s and at an electric field of 1000 V/m, with increase in SDS concentration from 5 kg/m^3 to 30 kg/m^3, gel layer thickness and (R_g/R_m) increases from 4.66 μm to 25.8 μm and from 0.13 to 0.72, respectively.

4.5.2. Electric Field Assisted MEUF of SDS Solution Containing of Methylene Blue Dye

4.5.2.1. Effect of Surfactant to Dye Concentration Ratio

The effect of feed SDS to MB concentration ratio on the permeate flux and retention of MB is demonstrated in Figure 4.11. For a fixed MB concentration of 0.05 kg/m^3, the concentration ratio of SDS to MB is varied from 100 to 600. From the figure, it may be observed that the retention of MB increases with increase in SDS to MB ratio. For example, at zero electric field, a cross flow velocity of 0.09 m/s and a pressure of 360 kPa, with increase in SDS to MB ratio from 100 to 400, the retention of MB increases from 82% to 99%. No significant improvement of retention of MB is noticed on further increase in surfactant to dye ratio. For example, the retention of MB is about 99.0% when SDS to MB ratio increases to 600. Therefore, the surfactant to dye concentration ratio is optimum at 400 and the subsequent MEUF experiments are carried out at this ratio. As discussed earlier, it is clear from the figure that permeate flux decreases with increase in SDS concentration in the feed. At zero electric field, at a pressure of 360 kPa and cross flow velocity of 0.09 m/s, with increase in SDS to MB ratio (i.e., SDS concentration increases from 5 to 30 kg/m^3, MB concentration remains unchanged), permeate flux decreases from 50.4 L/m^2 h to 31.3 L/m^2 h. Similar trends are also observed in presence of electric field. For example, at an electric field of 1000 V/m, with increase in SDS to MB ratio (i.e., SDS concentration increases from 5 to 30 kg/m^3), permeate flux decreases from 61.2 L/m^2 h to 39.6 L/m^2 h.

The effect of methylelene blue (MB) concentration in the feed mixture of SDS and MB on the retention of MB and on permeate flux for different electric field is described in Figure 4.12. In this figure, SDS to MB concentration ratio is shown to vary by increasing the MB concentration from 0.01 kg/m^3 to 0.05 kg/m^3, keeping SDS concentration fixed at 20 kg/m^3. From the figure it is observed that the retention of MB is almost invariant with MB concentration. Similarly, the variation of permeate flux is insignificant as concentration of MB increases. The results shown in Figure 4.11 and 4.12 suggested that the concentration of SDS, not that of MB is the controlling factor in detecting the permeate flux of MEUF of this system. Similar trends are also observed in presence of electric field.

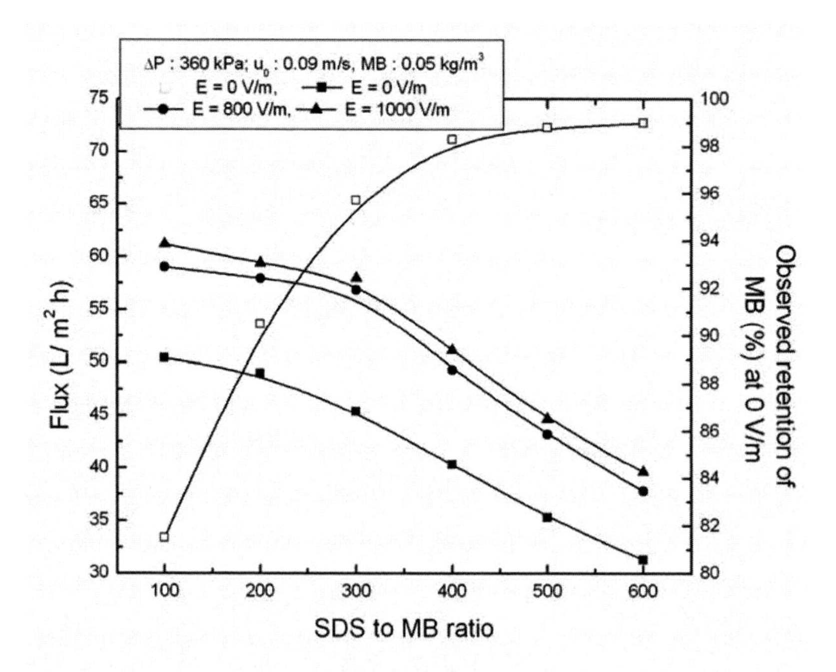

Figure 4.11. Variation of permeate flux and observed retention of MB with SDS to MB ratio for different electric field at a fixed pressure of 360 kPa and a cross flow velocity of 0.09 m/s.

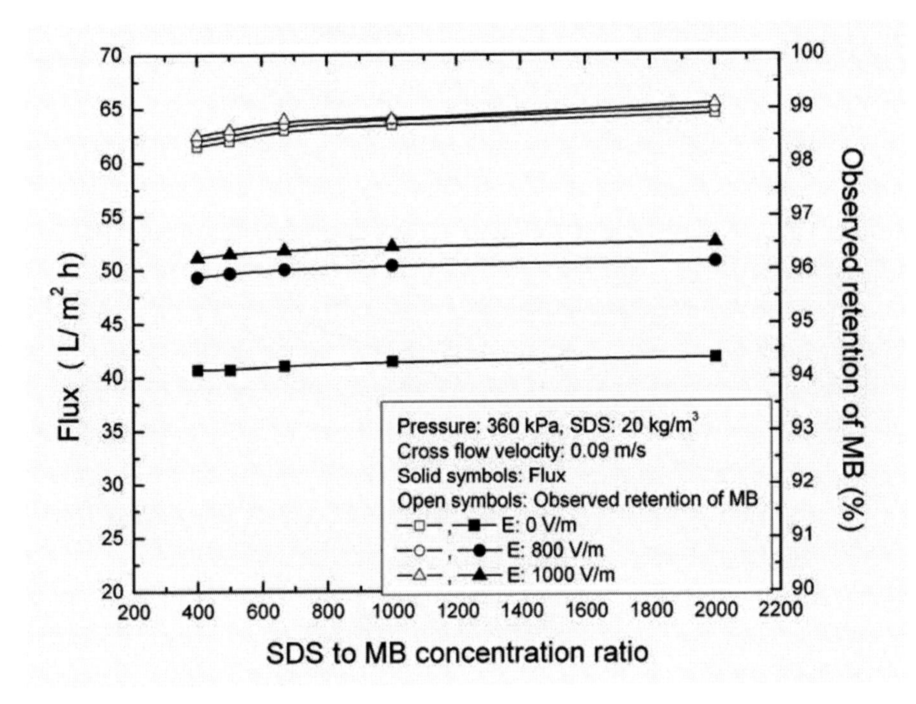

Figure 4.12. Effect of MB feed concentration on observed retention of MB and permeate flux at different electric field.

4.5.2.2. Effect of Electric Field

Figure 4.13 demonstrates the effect of external d.c. electric field on permeate flux and on observed retention of MB. It is clear from the figure that permeate flux increases with increase in electric field. As discussed earlier, with increase in electric field gel layer formation of micelle aggregates on the membrane surface is restricted due to electrophoretic movement of SDS micelle away from the membrane surface leading to a decrease in gel layer thickness. This results in an increase in permeate flux. For example, the permeate flux increases 26% with an increase in electric field from 0 to 1000 V/m, for a fixed SDS to MB ratio of 400, pressure of 360 kPa and cross-flow velocity of 0.09 m/s. It is further observed from the figure that the effect of electric field on the retention of MB is not significant. For example, with the same increment of electric field, the retention of MB remains invariant at 99%. This observation indicates that the retention of dye is entirely dictated by its solubilization only within the micelles.

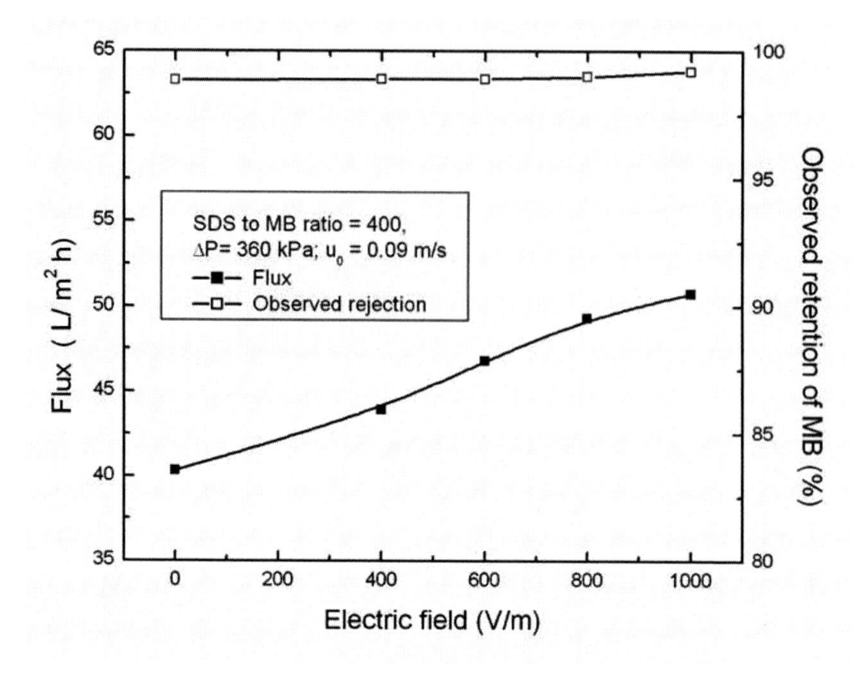

Figure 4.13. Variation of permeate flux and observed retention of MB with electric field at a SDS to MB concentration ratio of 400, a pressure of 360 kPa and a cross flow velocity of 0.09 m/s.

4.5.2.3. Effect of Cross Flow Velocity

The influence of cross flow velocity on the permeate flux and on observed retention are shown in Figure 4.14 for different SDS to MB ratio. This ratio is varied by changing the MB concentration at a fixed SDS concentration of 20 kg/m^3. With increase in cross flow velocity, the retention of MB is insignificant indicating cross-flow velocity does not have any impact on the retention of MB. However, increase in cross flow velocity improves the permeate flux. This is attributed to the facts that increase in cross flow velocity increases the sweeping effect on the micelle gel layer over the membrane surface. This restricts the deposition of micelles on the membrane surface; further it facilitates backward diffusion from the surface to the bulk. Hence, with increase in cross flow velocity, the gel layer thicknesses, and consequently resistance to solvent flow, decreases, thereby augmenting the permeate flux. For example, at 1000 V/m and SDS to MB concentration ratio 400, with increase in cross flow velocity from 0.09 m/s to 0.18 m/s, the permeate flux increase by about 16% and for pure SDS solution, with the same increment of cross flow velocity, the flux is enhanced by about 15%.

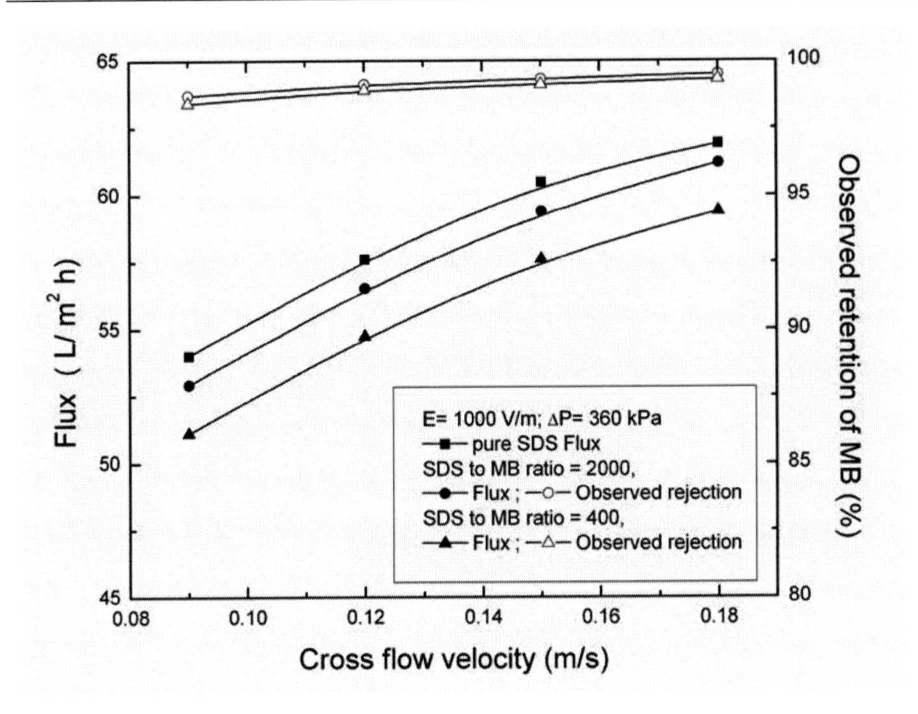

Figure 4.14. Variation of permeate flux and observed retention of MB with cross flow velocity at a SDS to MB concentration ratio of 2000, a pressure of 360 kPa and an electric field of 1000 V/m.

4.5.2.4. Effect of Pressure

Figure 4.15 demonstrates the variation of transmembrane pressure on the permeate flux and on the retention of MB. The symbols are the experimental flux data and solid lines are the model calculation. It is observed from the figure that the retention of MB is almost invariant with pressure. The variation of permeate flux with pressure shows the similar trend as observed in ultrafiltration of pure SDS solution (Figure 4.5). For SDS to MB concentration ratio of 400, electric field of 800 V/m and a cross flow velocity of 0.09 m/s the limiting pressure and limiting flux as discussed earlier are estimated to be 571 kPa and 60 L/m^2 h respectively. Keeping other operating conditions unchanged, at 0.15 m/s, the corresponding values are 617 kPa and 70 l/m^2 h. Furthermore, with increase in pressure, permeate flux increases. Experimental results show that for SDS to MB ratio of 400, at 800 V/m and 0.15 m/s, with increase in operating pressure from 220 kPa to 635 kPa, the permeate flux increases from 41 L/m^2 h to 70 L/m^2 h and the observed retention of MB decreases marginally from 99% to 98%.

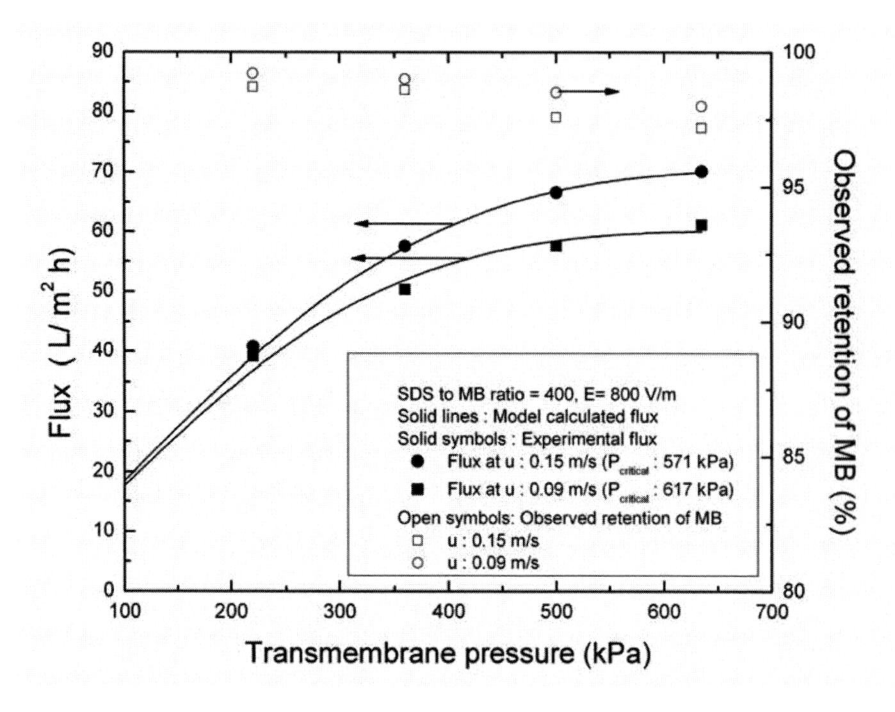

Figure 4.15. Variation of permeate flux and observed retention of MB with transmembrane pressure for different cross flow velocity at a fixed SDS to MB concentration ratio of 400 and at an electric field of 800 V/m.

4.6. Conclusion

One of the major areas of concern in the MEUF is the flux decline due to concentration polarization followed by gel layer formation of micelle aggregates over the membrane surface. In this chapter, the results of electric field assisted micellar enhanced ultrafiltration of methylene blue dye using SDS solutions carried out in a specially designed cross flow ultrafiltration module are presented. Under the application of an external d.c. electric field across the membrane, gel layer formation is restricted leading to an enhancement of permeate flux. From experimental results, the optimum surfactant to dye concentration is found to be 400. It may be concluded from the experimental results that permeate flux increases with electric field, cross flow velocity, transmembrane pressure and decreases with surfactant concentration. The retention of solute shows a strong function of surfactant concentration. Electric field, cross flow velocity and transmembrane pressure are shown to have very little effect on solute retention. Moreover, electric field assisted MEUF is found to be equally effective as conventional MEUF in terms of solute retention but with enhanced permeate flux. External d.c. electric field can be applied to increase the permeate flux during micellar enhanced ultrafiltration of organic solutes which are stable under electric field. Enhancement of permeate flux by applying pressure is possible up to a certain pressure called limiting pressure. In this chapter, the expressions for prediction of the limiting pressure and limiting flux are proposed which will be useful for the design of membrane separation modules.

REFERENCES

[1] C. Teng, V. Chen, Nanofiltration of textile wastewater for water reuse, *Desalination* 143 (2002) 11-20.

[2] L. Tinghui, T. Matsuura, S. Sourirajan, Membrane materials and average pore size on reverse osmosis separation of dyes, *Ind. Eng. Chem. Prod. Res. Dev.* 22 (1983) 77-87.

[3] Y. Xu, R. Lebrun, P.J. Gallo, P. Blond, Treatment of textile dye effluent by nanofiltration, *Sep. Sci. Technol.* 34 (1999) 2501-2519.

[4] J.F. Scamehorn, J.H. Harwell (Eds.), *Surfactant Based Separation processes,* Surfactant Science Series, Marcel-Dekker Inc., New York, 1989

[5] R.O. Dunn Jr., J.F. Scamehorn, S.D. Cristian, Use of micellar enhanced ultrafiltration to remove dissolved organics from aqueous stream, *Sep. Sci. Technol.* 20 (1985) 257-284.

[6] M. Syamal, S. De, P.K. Bhattacharya, Phenol solubilization by cetylpyridinium chloride micelles in micellar enhanced ultrafiltration, *J Membr. Sci.* 137 (1995) 99-107.

[7] M.K. Purkait, S. DasGupta, S. De, Removal of dye from waste water using micellar enhanced ultrafiltration and recovery of surfactant, *Sep. Sci. Technol.* 37 (2004) 81-92.

[8] S.R. Jadav, N. Verma, A. Sharma, P.K. Bhattacharya, Flux and retention analysis during micellar enhanced ultrafiltration for the removal of phenol and aniline, *Sep. Sci. Technol.* 24 (2001) 541-557.

[9] A.L. Ahmad, S.W. Puasa, M.M.D. Zulkali, Micellar-enhanced ultrafiltration for removal of reactive dyes from an aqueous solution, *Desalination* 191 (2006) 153-161

[10] N. Zaghbani, A. Hafiana, M. Dhabhi, Separation of methylene blue from aqueous solution by micellar enhanced ultrafiltration, *Sep. Sci. Technol.* 55 (2007) 117-124

[11] C. Fesi, L. Gzara, M. Dhabhi, Treatment of textile effluents by membrane technologies, *Desalination* 185 (2005) 399-409.

[12] B. Van der Bruggen, B. Daems, D. Wilms, C. Vandecasteele, Mechanisms of retention and flux decline for the nanofiltration of dye baths from the textile industry, *Sep. Sci. Technol.* 22-23 (2001) 519-528.

[13] M. Bielska, K. Prochaaska, Dyes separation by means of cross-flow ultrafiltration of micellar solutions, Dyes and pigments 74 (2007) 410-415

[14] J.F. Rathman, J.F. Scamehorn, Counter binding on mixed micelles, *J. Phys. Chem.* 88 (1984) 5807-5816

[15] W.F. Blatt, A. Dravid, A.S. Michaels, L. Nelson, Solute polarization and cake formation in membrane ultrafiltration: causes, consequences and control techniques, in: J.E. Flinn (Ed.), *Membrane Science and Technology,* Plenum Press, New York, 1970.

[16] G. Akay, R.J. Wakeman, Electric field intensification of surfactant mediated separation processes, *Trans. Inst. Chem. Eng.* 74 (1996) 517–525.

[17] Roessler, D. Crettenand, O. Dossenbach, W. Marte, P. Rys, Direct electrochemical reduction of indigo, *Electrochim. Acta.* 47(2002) 1989-1995.

[18] V. Golob, A. Vinder, M. Simonic, Efficiency of the coagulation/f locculation method for the treatment of dyebath effluents, *Dyes pigments* 67 (2005) 93-97.

[19] G.M. Walker, L.R. Weatherley, Biodegradation and biosorption of anthraquinone dye, *Environ. Pollut.* 108 (2000) 219-223.

[20] Wang, A. Yediler, D. Lienert, Z. Wang, A. Kettrup, Ozonation of an azo dye C.I. Remozol Black5 and toxicological assessment of its oxidation products, *Chemosphere* 52 (7) (2003) 1225-1232.

[21] S. Wang, Z.H. Zhu, A. Coomes, F. Haghseresht, G.Q. Lu, The physical and surface chemical characteristics of activated carbons and the adsorption of methylene blue from waste water, *J. Colloid Interface Sci.* 284 (2005) 440-446.

[22] L. Song, A new model for the calculation of the limiting flux in ultrafiltration, *J. Membr. Sci.* 144 (1998) 173-185.

[23] J.S. Shen, R.F. Probstein, On the prediction of limiting flux in laminar ultrafiltration of macromolecular solution, *Ind. Eng. Chem., Fundam.* 16(4) (1977) 459-465.

[24] J.F. Scamehorn, J.H. Harwell (Eds.), Surfactant Based Separation Processes, *Surfactant Science Series,* Marcel-Dekker Inc., New York, 1989.

[25] S.R. Epton, A new method for the rapid titrimetric analysis of sodium alkyl sulphates and related compounds, *Trans. Faraday Soc.* 44 (1948) 226–230.

[26] J. B. Hayter and J. Penfold, Determination of micelle structure and charge by neutron small-angle scattering, *Colloid Polym. Sci* 261 (1983) 1022-1030.

[27] S. De, P.K. Bhattacharya, Modeling of ultrafiltration process for a two – component aqueous solution of low and high (gel-forming) molecular weight solutes, *J. Membr. Sci.* 136 (1997) 57-69.

[28] Y. Moroi, Mass action model of micelle formation: Its application to sodium dodecyl sulphate solution, *J. Colloid Interface Sci.* 122 (1988) 308-314.

[29] S.D. Christian, J.F. Scamehorn, Use of micellar enhanced ultrafiltration to remove dissolved organics from aqueous streams, in: J.F. Scamehorn (Ed.), *Surfactant Based Separation Process,* Marcel Dekker, New York, 1989.

[30] P. Bacchin, A possible link between critical and limiting flux for colloidal systems: concentration of critical deposit formation along a membrane, *J. Membr. Sci.* 228 (2004) 237-241.

INDEX

B

C

D

E

F

G

H

I

N

O

P

T

U

V

W

X

Y

Z